Arup's Tall Buildings in Asia

Through a series of detailed case studies from East Asia, Arup, one of the global leaders in tall building design, presents the latest developments in the field to inspire more innovative and sustainable ideas in tall building design and engineering.

This book exhibits the key design aspects of tall buildings in 20 case studies, from China, Singapore, Hong Kong, Vietnam and Japan. Chapters cover design and construction, safety concerns, sustainability strategies, BIM and optimisation solutions, and include contributions from the actual project engineers. The projects chosen are not the tallest buildings, but all of them have been selected for their significant engineering insights and values. Arup's engineers explain the design principles, and how they overcame various design constraints and challenges, while exceeding their clients' expectations.

Unique examples include:

- the design and application of a hybrid outrigger system in the Raffles City Chongqing project
- the challenges encountered in the construction of the CCTV Headquarters, Beijing
- as well as Tianjin's Goldin Finance 117 Tower, Ho Chi Minh City's Vincom Landmark 81, the China Resources Headquarters, Ping An IFC, Tokyo's Nicolas G Hayek Center and the Shanghai World Financial Centre.

These varied and complex case studies draw on multi-disciplinary design and engineering challenges which make this book essential reading for architects, structural engineers, project managers and researchers of high-rise buildings. The book also provides a useful reference for and link between practitioners in the industry, academia and engineering students.

Goman Wai-Ming Ho is an Arup Fellow. He has extensive experience in multi-disciplinary and mega scale and tall building projects, especially in the East Asia Region. He is currently the global leader of the Tall Building Skills Network of Arup. He joined Arup in 1992 after completing his PhD at Hong Kong Polytechnic University. He has been involved in projects across the East Asia Region including: Hong Kong, Japan, Korea, Mainland China, Myanmar, the Philippines, Taiwan and Vietnam. He is a registered Chartered Engineer in the UK, Registered Professional Engineer in Hong Kong and First Class Registered Structural Engineer in the People's Republic of China.

Arup's Tall Buildings in Asia

Stories behind the Storeys

Chief Editor
Goman Wai-Ming Ho

Contributing Editor
Ruby Kitching

Editors
Anita Siu
Christina Yang

Routledge
Taylor & Francis Group
LONDON AND NEW YORK

First published 2018
by Routledge
2 Park Square, Milton Park, Abingdon, Oxon OX14 4RN

and by Routledge
711 Third Avenue, New York, NY 10017

Routledge is an imprint of the Taylor & Francis Group, an informa business

British Library Cataloguing-in-Publication Data
A catalogue record for this book is available from the British Library

Library of Congress Cataloging-in-Publication Data
Names: Ho, Goman Wai-Ming, editor.
Title: Arup's tall buildings in Asia : stories behind the storeys / Goman Wai-Ming Ho [editor].
Description: New York : Routledge, 2017. | Includes bibliographical references and index.
Identifiers: LCCN 2017016560 | ISBN 9781138058736 (hardback : alk. paper) | ISBN 9781315164021 (ebook : alk. paper)
Subjects: LCSH: Arup Group Ltd. | Tall buildings–Asia. | Tall buildings--Design and construction--Case studies.
Classification: LCC TA217.O94 A78 2017 | DDC 690/.383095–dc23
LC record available at https://lccn.loc.gov/2017016560

ISBN: 978-1-138-05873-6 (hbk)
ISBN: 978-1-315-16402-1 (ebk)

Typeset in Univers
by Saxon Graphics Ltd, Derby

Contents

About ARUP

Arup is the creative force at the heart of many of the world's most prominent projects in the built environment and across industry.

Founded in 1946 with an initial focus on structural engineering, Arup has since grown into a truly multi-disciplinary organisation. Today we have 13,000 designers, engineers, planners and consultants based in 85 offices across 35 countries.

Through its work, Arup is making a positive impact on people's lives – from transport infrastructure that links people with opportunities to galleries and concert halls that offer a sense of delight.

This book covers only part of Arup's diverse work, giving a peek into its ingredients to develop better solutions: the Total Design approach. For Arup, this means forging close collaboration between the client, architect and engineer as well as bringing together Arup people from a wide range of disciplines to look beyond their own specialisms. The joined-up thinking enables the firm to see the big picture, foster innovation and help owners and architects to realise ambitious structures that are commercially viable, operationally efficient and socially engaging.

This unique approach is firmly underpinned by Arup University (AU), the firm's research, learning and knowledge-sharing programme. Unlike conventional corporate universities, AU is designed to connect and empower both Arup people and the wider industry to explore fresh ideas, shape new skills and drive technical excellence to tackle some of the big issues of today's built environment whilst meeting our clients' aspirations.

Presented by AU, this book is a compelling statement of how Arup continues to shape a better world with a strong commitment to innovation, technical prowess and sharing for greater impact.

Preface

Soaring hundreds of metres into the sky, tall buildings are awe-inspiring structures that achieve iconic status by overcoming some of nature's greatest obstacles. They reflect a nation's confidence and hope for economic success, while still carrying out the essential functions that allow urban districts to thrive.

Accommodating thousands of people on a small footprint, many of the tallest buildings described in this book operate like highly efficient vertical cities. The need for them in Asia comes as a result of rapid urbanisation. The trend is expected to continue and, with it, the demand for high-density developments and building tall.

Arup is a global engineering consultancy which has been working in Asia for over 40 years and has played a part in the region's transformation, designing buildings that are tall, safe, efficient, sustainable and also fabulous and intriguing to look at. Using the latest thinking and digital tools, the most incredible structures have been conceived, despite architecture and building functions becoming more complex and the effect of climate change creating additional pressures. Increasing demands have also been placed on occupant comfort and vertical transit. These challenges drive engineers to come up with innovation upon innovation to allow these edifices to rise up from the ground, undefeated.

The future of tall buildings, as dictated by market needs, can be to scale heights much greater than described in this book, but will also have to rely on overcoming their many challenges. Finding ways to make structural materials perform more efficiently or for systems – whether related to structure or services, safety or the environment – to work more effectively, are some of the most enchanting aspects of tall building design. This is where radical ideas can still find a home and be implemented to the sheer joy of the engineer.

Schemes that were once considered impossible to build are now operational and serve as inspiration for the next generation of tall buildings whose height may only be part of the story that makes them interesting. It is the buildings in which the boundaries of conventional architectural, structural and services design have been crossed that will become the true classics of their time.

With this in mind, each chapter of this book expresses the qualities that make these buildings special and gives examples of Arup's ingenuity that has made many so successful – whether it is the structural gymnastics that holds them up, the rationale that allows thousands of façade panels to be fabricated and installed with millimetre accuracy, or the services strategy which makes a positive contribution to the environment.

Arup's longstanding experience and its will and determination to investigate and test where there is no precedent continue to keep the firm at the forefront of engineering design; its "Total Design" philosophy – a multi-disciplinary approach to working – being key to developing the best possible, holistic solutions.

With Asia being home to so many extraordinary tall buildings, each with their own inimitable style, the region offers many examples of how the genre is moving forward. Arup's design stories of tall buildings in Asia are being shared now to invite an exchange of ideas across the global building community and to make sure the sky really is the limit when it comes to tall building design.

Goman Wai-Ming Ho
Arup Fellow, Tall Buildings Skill Leader

Foreword

With an exponential growth of urban population, the tall building has become the most dominant symbol of cities, economic progress and technological advancement. Nowhere is the phenomenon more pronounced than in Asia where a new group of skyscrapers is emerging – multi-function, unprecedented height with a strong ambition for form and performance.

Realising these ambitious towers, from the latest addition to Beijing's skyline, China Zun, to the jewel of Ho Chi Minh City, Landmark 81, has cemented Arup's reputation for being the trendsetter in building engineering. This book captures some of the best examples of Arup's work over the past decade, through which the reader can follow the design process, trace the evolution of the design and have a glimpse into the technical excellence embedded within.

More importantly, the Arup approach of "Total Design" is well illustrated in the design process of tall buildings – mini vertical cities that rely on a complex web of systems and services to thrive. We have divided the book into different chapters – each shining a light on solutions for a particular challenge as a result of the synergy created by aligning our vast skillset and engineering disciplines.

Of course, Total Design is more than integrating expertise from within; it's an ingenious combination of engineering, architecture and construction technology. Each of the showcase projects is a triumph of creative collaboration and partnership; we could not have done it without our wonderful clients and collaborators.

This creative collaboration also reinforces our tradition of research for innovation – an integral part of many projects and fundamental to our pursuit of excellence. The ideas we pioneered in these projects have formed the key part of many tall building projects worldwide. Exploring new tools and techniques keeps us at the forefront of the market, and working together with external parties ensures that our research creates value for our clients, the industry and society as a whole.

This book is published as another endeavour to forge collaboration through knowledge sharing which is at the heart of spawning innovation and progress. We are proud to share these stories and look forward to inspiring more ideas to shape a better future.

Michael Kwok
East Asia Region Chairman, Arup

Foreword

Tall buildings are not only a product of cities in human society, but also a necessary outcome of the diversified needs and challenging limits that are intrinsic to human beings. The development of tall buildings has become the solution to the scarcity of land resources in modern cities and serve as an antidote to the problems of rising populations, huge consumption of energy, and climate change. Tall buildings are also sophisticated products of the times, resulting from the development and integrated application of various innovative technologies. As indicators of the cultural quality of cities, tall buildings are attuned to the harmony of a city's environment to preserve its ecology. However, tall buildings bear enormous responsibilities of withstanding massive earthquakes and winds, as well as the challenge of timely evacuation of their occupants during emergencies such as fires. The costs are also exorbitant for tall buildings which are used for long periods of time. Therefore, the design of tall buildings should consider the comfort, safety, and sustainability of living in them.

With the shift in the global economic centres and the rise of Asian countries, urbanisation is spreading all over Asia. The development of tall buildings in Asian countries is inevitable. On the one hand, today's digital technologies, communication networks, new materials and high technology have promoted the development of tall buildings. On the other hand, they have presented challenges to designers of tall buildings.

Dr Goman Wai-Ming Ho and his colleagues at Arup have embraced these challenges. With more than 40 years' experience in Asia, Arup has played an important part in the region's urbanisation. It has combined local insights with global expertise to deliver a wide range of landmark projects. This book, *Arup's Tall Buildings in Asia: Stories behind the Storeys*, presents its multi-disciplinary skills and rich experience in designing and constructing striking landmarks across the built environment in Asia.

The Faculty of Construction and Environment (FCE) of The Hong Kong Polytechnic University has had a long relationship with Arup. In addition to nurturing numerous engineers and managers for Arup, it has collaborated with Arup in research on tall buildings. Dr Goman Wai-Ming Ho is an alumnus and an adjunct professor of the Hong Kong Polytechnic University. This book will no doubt provide inspiration as well as practical guidance to students, professors, architects and engineers alike. Arup's expertise in tall buildings will continue to help the region grow beyond the sky's limit.

Prof. You-Lin Xu, FASCE, FEMI, FHKIE, FIStructE
Yim, Mak, Kwok & Chung Professor in Smart Structures
Dean of Faculty of Construction and Environment
The Hong Kong Polytechnic University

Chapter 1

Ambitiously tall

The fascination of building begins as a child with stacking wooden blocks to create a tower. The thrill and excitement of those early days is amplified for structural engineers designing tall buildings where the need to reach greater heights (and make sure the blocks do not tumble down) is played out on a much larger scale.

A growing trend for more and more unconventional architectural forms is creating an equal mix of opportunity and constraint for engineering designers where the most challenging schemes are not just extremely tall, but may also hold other complex structures aloft. Working closely with architects to develop structural solutions, these framing systems can be bold and profoundly influence a building's identity, or be completely concealed from view.

The tall buildings described in this book make a statement on the landscape: symbolic of power for a city and pride for its people. In the role of structural engineer, Arup uses many different methods to allow these urban icons to reach their summits safely and efficiently, often involving specialist computer analysis as well as good, old-fashioned human ingenuity. Structural systems made up of outriggers and bearings, belt trusses and bracing, mega columns and pioneering vertical stressing lift an architect's dream off the drawing board and into (steel and) concrete reality.

1a

© Safdie Architects/CapitaLand (China) Investment Co. Ltd

Case study 1a

Raffles City Chongqing
Chongqing, China

CONTRIBUTORS: ANDREW LUONG, PENNY CHEUNG + JIE LIU

Client:	CapitaLand Ltd
Design architect:	Moshe Safdie and Associates; P & T Group, Chongqing Architecture and Design Institute
Arup's role:	Structural, civil and fire safety engineering plus sustainability services for all design stages
Year of completion:	2018
Height:	Eight towers: T1 & T6 ~250m T2, T3S, T4S & T5 – ~270m (including conservatory) T3N & T4N – 350m
Number of storeys:	46 to 79 floors above ground with 3 floors basement
Gross floor area:	1,129,423m^2
Use:	Mixed
Main contractor:	Tender A for China State Construction Engineering Corporation (CSCEC) Third Bureau. Tender B for CSCEC Eighth Bureau

Chongqing is a rapidly growing city in western China at the confluence of the Yangtze and Jialing rivers. An important trading centre connecting the east and west of China, the municipality (under direct central government control since 1997) is undergoing expansion and regeneration.

Celebrating prosperity and heralding the city's future successes is the iconic Raffles City Chongqing megasculpture – an eight-tower mixed-use development made up of curved, slender buildings which mimic the billowing sails of ancient ships which once passed along its great rivers.

The tallest two towers to the north of the development are 355m tall and form the front of the "ship". They are connected to a bank of four 265m tall towers by two high-level link bridges (a 20m wide pedestrian bridge and a 5m wide fire escape route). Two further 235m tall freestanding towers complete the mixed-use development, which sits on a nine-storey retail podium.

On top of the bank of four towers is a 280m long dazzling steel and glass conservatory that houses the hotel lobby, an observatory, restaurant, pool and

1a1
Raffles City Chongqing: an eight-tower megasculpture with stunning rooftop conservatory
© Safdie Architects/ CapitaLand (China) Investment Co Ltd

spa. So many aspects of this development – its curves, height and slenderness – push the boundaries of tall building design, but none more so than the conservatory. While much engineering ingenuity went into each tower at Raffles City Chongqing, particularly in developing the structural system for the tallest north towers, the conservatory presented the greatest complexities.

Engineers have been designing tall buildings and long span structures for a long time. But designing a combination of the two – a tall, inhabited structure and a long spanning one at high level – is a new proposition and a unique approach was needed to ensure Raffles City Chongqing would be structurally and economically feasible.

Understanding how the four towers and the conservatory would interact with each other was the main point to consider – should the conservatory be physically connected to the towers and move when they moved under high winds, seismic activity, settlement or thermal response? Or should it be isolated from the towers so that the conservatory would be unaffected by any tower movement? And in this isolated scenario, was it possible for a conservatory to float 250m high above the ground?

At the start of the design process, there was no obvious solution, but engineers considered that, since Chongqing is an area of comparatively low seismic activity, the most economic design should be based on a "normal" rather than "extreme" loading scenario. In addition, the design would need to incorporate some means of withstanding the most extreme conditions arising from an earthquake with the probability of occurring once every 1,600 to 2,500 years.

Although the two tallest towers are linked to the conservatory towers via two bridges, it was considered early on that these connections should each incorporate a 1.8m flexible zone adjacent to the north towers to absorb the substantial differential movement which would occur across the towers. This meant that the four towers and conservatory could work as a separate structural system to the north towers.

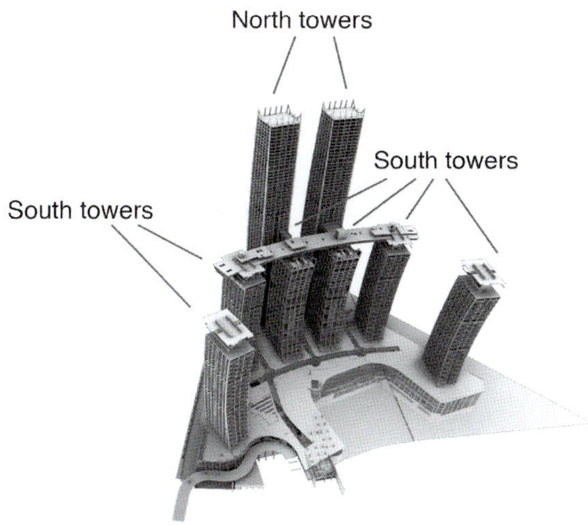

1a2
Raffles City Chongqing: a mixed-use development
© Arup

How to make a conservatory float

Three scenarios were considered to connect the conservatory to the towers. For comparison, a simple base case for the scheme with a monolithic arrangement was initially explored (see Figure 1a3). Here, with the conservatory rigidly connected to the towers, the forces developed in the structure would be large, requiring large structural components and bulky detailing which would have significant cost implications and encroach on lettable space.

The second scenario considered splitting the conservatory into sections so that each tower supported one portion. Movement joints in the conservatory would allow each portion to move relative to one another while still being fixed to the tower. Further analysis of this option revealed that provision for up to 3m of relative movement would be needed for this system to work. A movement joint which could accommodate this amount of movement would be unsightly in the sleek conservatory, as well as impose too many restrictions on how the space could be used. Constructing the hammerhead towers would also take longer and, hence, be more expensive, so this solution was also ruled out.

The third scenario explored the use of bearings, which would have the effect of reducing the magnitude of the forces transferred to the conservatory from the towers during earthquakes. Connecting the conservatory and four towers dynamically would lead to a lighter structure, which appealed to the

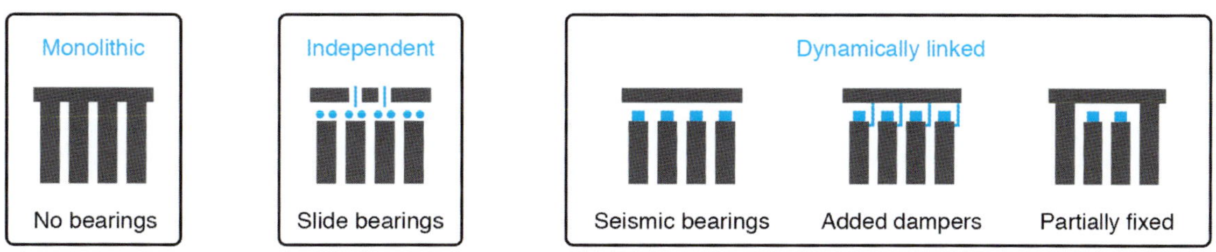

1a3
Exploring the connectivity between the conservatory and towers
© Arup

architect and would have the added benefit of being much less expensive than the base case. The impact of using bearings on all towers or just two and allowing the other two towers to be rigidly connected to the conservatory was investigated to determine where efficiencies could be made. To limit movement experienced by the conservatory and towers during an earthquake and to reduce the size of bearing required, dampers were also considered. The expectation was that a lighter structure would become feasible if a damping mechanism was incorporated, leading to a reduction in material costs.

The combination of fixed and isolated conservatory–tower connections was quickly ruled out due to problems arising from the fixed connections (as described for the monolithic case) and high material costs, which left one possible solution – to use bearings on top of all four towers.

In all, 900 linear and non-linear analyses were carried out to explore scenario three, each with a typical run time of 30 hours, taking approximately 27,000 hours in total.

Friction pendulum versus lead rubber bearings

Friction pendulum bearings (FPBs) and lead rubber bearings (LRBs) were considered for the tower–conservatory connections. Both would work by being able to resist a certain amount of movement during a low level earthquake or high winds, but have the ability to kick into action during more extreme loading conditions. The LRB option was eventually ruled out as they could not be designed to resist the highest seismic forces expected in Chongqing and would also be difficult to replace.

FPBs, in contrast, would function well during high intensity earthquakes, where the pendulum mechanism would "swing" the conservatory back to its equilibrium position during a tremor. Even in the most extreme conditions, the bearings would not need to be replaced.

An FPB is made up of a lower concave dish and an articulated slider, which sits on a composite liner (Figure 1a4). Under normal load conditions, the FPB is fixed and the conservatory would not move and stresses arising from low seismic activity, wind and thermal movements would be resisted by the main structure. Under more moderate or severe earthquake conditions, the bearing mechanism – the slider – kicks into action, sliding against the composite liner and allowing the conservatory to move relative to the tower and "float". Isolating the conservatory in this way also protected its glass and slender steel truss structure from damage.

Fine-tuning the bearing design

Further analysis of FPB performance considered the optimum level of friction in the bearing and the effect of dampers. The higher the percentage of friction in an FPB, the greater the force to be resisted by the main structure before the pendulum mechanism is activated. Comparison of 2.5m and 1.5m diameter bearings with 3%

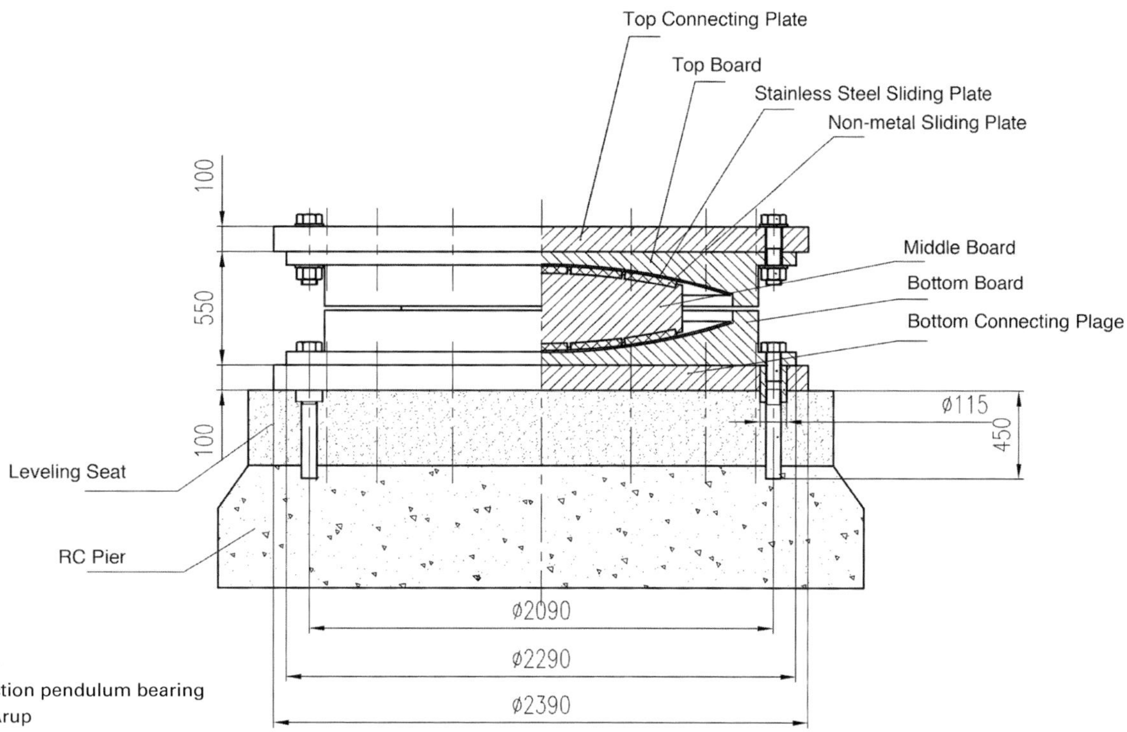

1a4
Friction pendulum bearing
© Arup

and 10% friction was investigated. The final optimised design uses six FPBs measuring 2m in diameter with between 4% and 7% friction in each of the four towers supporting the conservatory. Two extra bearings were included in the T4S tower (which includes a 20m wide cantilever bridge link to T4N).

Further analysis demonstrated that dampers would be beneficial in dissipating earthquake energy so that large forces, as well as the relative movement between conservatory and supporting towers, were not transferred into the conservatory structure. This solution, compared to the base case, further reduced the amount of steelwork in the conservatory.

A tale of two towers

The two north towers in Raffles City Chongqing are 355m tall and just 38m by 38m in plan, resulting in a slenderness ratio of 9.4 (compared to a slenderness ratio less than 8 for most super high-rises). These super slender buildings require special treatment to be able to withstand the high winds and earthquake activity prevalent in the region. To achieve the level of robustness required, tall, slender buildings usually use a system of mega columns and a braced mega frame. But as these towers would be used for hotel apartments as well as residential, the architect did not want bulky mega frame bracing obstructing views.

This was not so pertinent for the other six towers, which were subject to different architectural constraints due to their different occupancy types. These

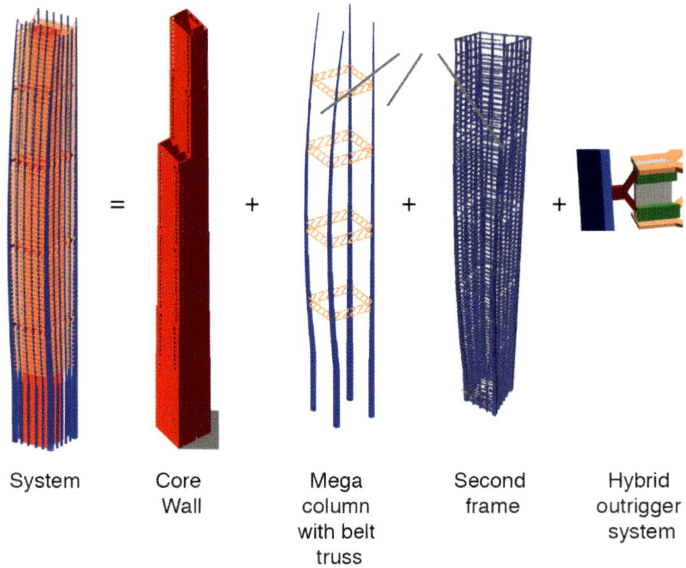

1a5
The structural system for the two north towers at Raffles City Chongqing, which preserve views out of the building
© Arup

shorter and stockier south towers accommodate a combination of shear walls, conventional outriggers and moment frames to achieve the stability required.

Nonetheless, to preserve clear views for residents of the taller, more slender north towers, an alternative approach was needed. The solution was to do away with the bulky braced mega frame and use mega columns at each corner of the building, four belt trusses spaced evenly along the height of the building and a discreet perimeter moment frame with outriggers. The main purpose of the outrigger, which reaches out from the concrete core to the corner mega column, like a skier using his arms and shoulders to hold onto the ski-poles, was to provide stability in an earthquake and during high winds (Figure 1a5). However, the "arm" component of the outrigger would need to be about 8m deep using concrete to resist the force of a high intensity earthquake and would significantly reduce the usable floor area in the building, making it economically unfeasible. Designed in steel, although the outrigger arm could be smaller, it would cost much more and take longer to construct.

To benefit from the cheaper cost of concrete and higher strength and ductility of steel, Arup developed a composite outrigger where the concrete portion could be constructed concurrently with the concrete core. The composite system would consist of a concrete portion connected to the core and a steel portion connected to a steel mega column. This steel component also incorporates an energy dissipating "fuse", giving it hybrid functionality suitable for normal and more extreme loading conditions (Figure 1a6).

During average wind and relatively low seismic activity (with a return period of 1 in 50 years), the composite outrigger provides the stiffness required to maintain stability with the fuse remaining intact. With a moderate (return period of 475) or severe earthquake (return period of 1,600 to 2,500 years), the fuse would yield first, dissipating the high forces in the building without the rest of the outrigger, mega column and core wall being damaged. The design considered

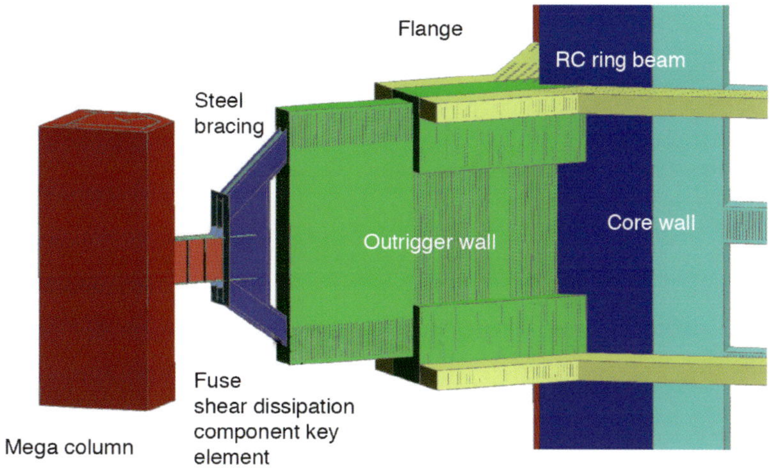

1a6
The hybrid composite outrigger
© Arup

1a7
Outrigger prototype testing
© Arup

that this steel "fuse" component could be cut out after yielding and replaced to return the building to its normal working mode.

In all, each tower uses four outriggers on each of its four refuge levels. This design led to approximately 50% cost savings compared to an all-steel outrigger solution. The innovation's other cost saving is attributed to faster construction for the composite outrigger compared to the pure steel or concrete versions. Had the outrigger been purely steel, then floor construction and all other activities on the tower would have had to wait one to two months while the steel outriggers for each refuge floor were installed. By using the composite outrigger, each floor's outriggers could be installed within the space of around two weeks.

The main components of the outrigger were carefully designed and tested using a prototype to validate the yielding mechanism. This involved specifying low strength steel for the fuse to ensure it would yield first and carefully detailing the other steel and concrete components to be more robust.

The composite hybrid outrigger addresses the unique challenges of designing slender towers in earthquake zones. The solution is a world first and has been patented by Arup. It opens up new possibilities for efficient tall building design in challenging environments.

Structural efficiency and innovations

At the heart of Raffles City Chongqing's design has been the desire to achieve tall, graceful towers and a sleek, sparkling jewel-like statement conservatory using the most appropriate structural philosophy, while applying innovative solutions for further efficiency gains. The bearing solution for the conservatory and the north towers' patented outrigger concept have together resulted in significant cost savings in the structure and delivered a solution which meets the architect's and client's aspirations for a landmark "skyscraper city" within a city.

1b
© P&T/Goldin Properties

Case study 1b

Tianjin Goldin Finance 117 Tower
Tianjin, China

CONTRIBUTOR: GOMAN WAI-MING HO

Client:	Tianjin Hi-Tech New Star Property Development Co Ltd
Design architect:	Palmer & Turner Consultants (Shanghai) Ltd
Associate architect:	East China Architectural Design & Research Institute Co Ltd
Arup's role:	Structural, geotechnical and wind engineer, façade consultant
Year of completion:	2018
Height:	597m
Number of storeys:	117
Gross floor area:	370,000m^2
Primary use:	Office and hotel
Main contractor:	China State Construction Engineering Corporation (CSCEC) Third Bureau

Located just 140km southeast of Beijing, Tianjin is a fast-emerging financial centre in northern China. It is one of four municipalities, along with Beijing, Shanghai and Chongqing, identified for growth by China's central government. Situated on the ancient Grand Canal which links the Yellow and Yangtze rivers, Tianjin was originally one of the oldest trading cities in the country.

Adjacent to the ancient city is a new Tianjin, favoured by the many multinationals operating in the country. And, like many other places that have attracted commerce and industry the world over, this new city needs a focus; a signature. This comes in the form of the Tianjin Goldin Finance 117 Tower, a 597m tall building expected to be the second tallest in China when it is completed in 2018. Taking its name from the developer and the number of floors in the building, the upper 20 levels will house a six-star hotel, while offices will occupy the remaining floors.

The tower is the jewel in the crown of the new 100,000m^2 Tianjin Goldin Metropolitan development, which includes shopping areas, luxury apartments, parks and leisure facilities. Its rooftop is covered by faceted glazing – literally looking like a diamond – and will accommodate a restaurant and observation deck for visitors to enjoy spectacular views across the city.

1b1
The jewel-topped Tianjin Goldin Finance 117 tower
© P&T/Goldin Properties

Affectionately known as the "magic stick", the tower is characterised by its slender profile, which is a 65m × 65m square at the base tapering to 45m along each side at roof level. Designing such a slim tower in an area of high seismic activity with the inherent problem of high wind loads on a tall building was challenging enough for engineers, but the building is also founded on soft ground, compounding the situation.

Coupled with the need to construct the building using local materials and labour within a tight construction programme and to a high specification with exemplar sustainability credentials, engineers had to pull out all the stops to get the project off the drawing board. Engineers had to develop performance-based design principles in consultation with existing codes and an expert review panel to justify the choice of structure. Steel and concrete have been used to their maximum potential to achieve an outstanding, structurally stiff, yet economical building in-keeping with the architect's vision. The structure also maximises floor space, satisfies fire performance requirements and facilitates large expanses of glazing to offer minimum obstruction to views across the city.

The structural components of this building comprise four steel mega columns at each corner of the building with eight single or double-storey transfer steel "belt" trusses distributed evenly along its height. Under severe seismic activity, the belt trusses are crucial in preventing progressive collapse. Visually, they also mark out each structural "zone". Together with perimeter steel cross-bracing and a central core, the structural system deals with all lateral earthquake and wind loading

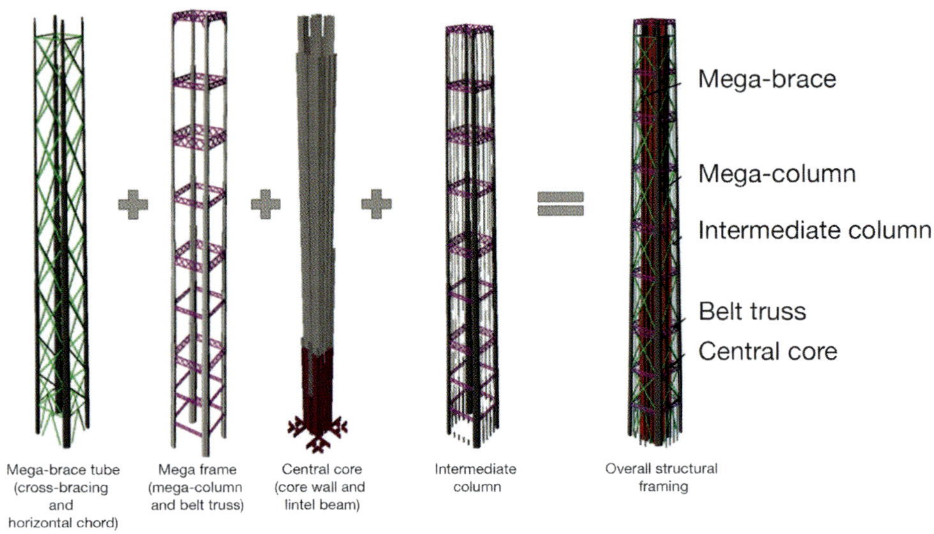

1b2
The structural frame
© Arup

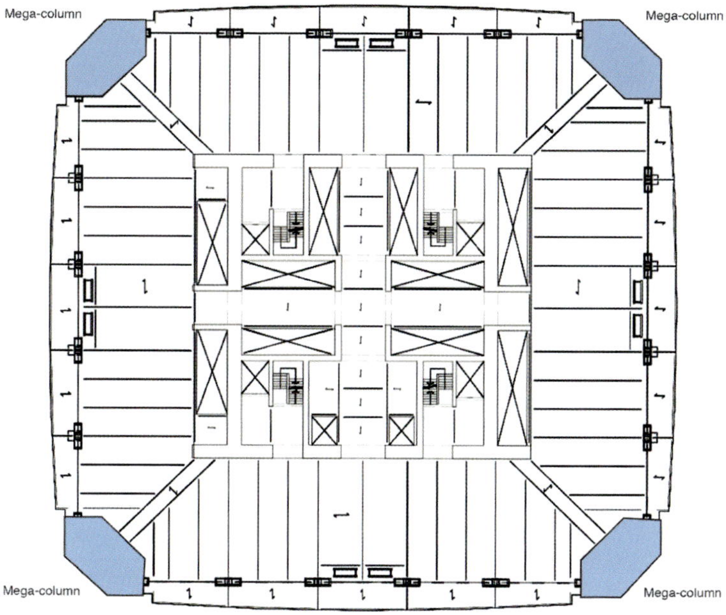

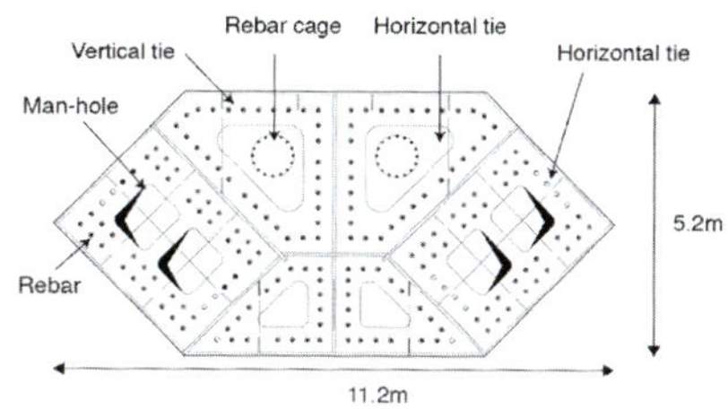

1b3
A typical floor plan showing location of mega columns (left) and cross section of a mega column (right)
© Arup

concerns as well as gravity loads. Floors are made up of simply supported steel beams and 120mm thick composite slabs on office floors (130mm thick on hotel floors).

The structural system transfers loads from each floor to perimeter columns within each structural zone to the belt truss, which in turn transfers loads to the corner mega columns. This arrangement greatly reduces the section size of the perimeter columns and creates an elegant structural arrangement without sacrificing lateral stiffness. Tianjin 117's robust central core and perimeter frame provide the extreme stiffness required for this "magic" stick to stand.

A structural solution with inner strength

Four mega columns connect to belt trusses and cross-bracing within each structural zone. Hexagonal in cross-section, the mega column cross-sectional area ranges from 45m^2 at the base of the tower to just 5.4m^2 at the top. Reducing the section size in line with structural requirements helps keep the overall weight of the building down to reduce loads on the foundations. This is a philosophy followed throughout this building's design, where possible. Although the mega columns obstruct the sought-after corner views, their location offers the most efficient and aesthetically acceptable solution to receive loads from the belt trusses.

To build the 120mm thick steel plate mega columns, which measure a maximum of 11.2m by 5.2m and span the full 597m height of the building, required a much smaller scale solution for accurate fabrication. This involved splitting the cross-section into multi-cellular, concrete-filled chambers which would also be easier to erect using tower cranes. Each 6.7m tall unit was detailed

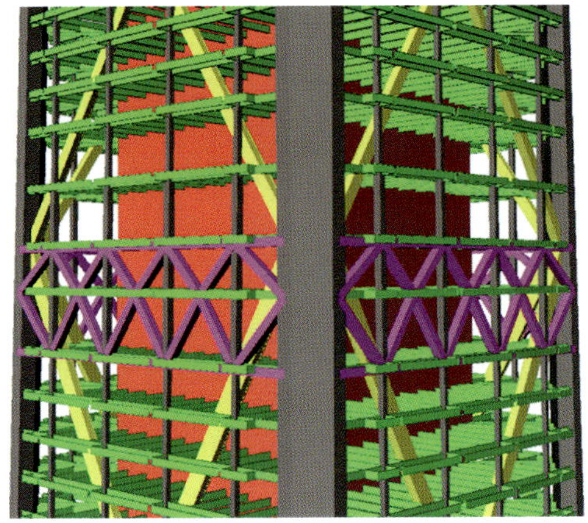

1b4
Belt truss at intermediate floor
Left: © Zhou Ruogu Architecture Photography Right: © Arup

in accordance with the specific structural requirements of its location to satisfy architectural, fabrication and erection needs, achieving the best overall economic and structural performance.

The six-sided polygon-shaped unit was fabricated in five components in the factory and then assembled adjacent to the tower. Additional checks were made on-site to ensure adjacent units fitted together correctly before being lifted into position. The chambers of each mega column unit were then filled with concrete to create an extremely stiff and robust composite mega column section. The interplay between mega column and belt truss is important for the tower's earthquake resistance – the truss resists most of the lateral load and transfers push and pull forces into the mega columns.

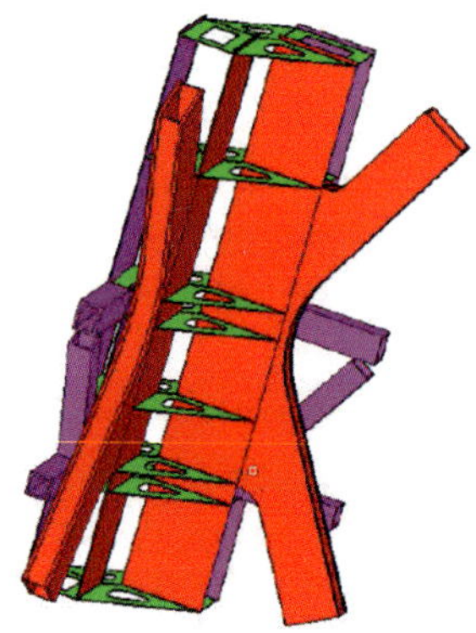

The core and cross-bracing

The core measures 34m × 32m at its base with an outer wall thickness of 1400mm and 600mm thick internal walls. Its footprint and wall thicknesses reduce higher up the building. The core adopts steel-reinforced concrete shear walls with embedded steel sections and, in the lower half of the building, also incorporates steel plates to prevent shear failure of the concrete walls in the event of a severe earthquake.

Cross-bracing throughout the building enhances the overall stiffness of the tower and ensures it satisfies seismic and wind loading code requirements. Around the entrance, a K-brace arrangement has been adopted to accommodate the building's main entrance.

The elegance of the Tianjin Goldin Finance 117 Tower comes from it being both extremely tall and extremely slim – an unusual combination in an earthquake zone – and creates a bold and dynamic statement for the new Tianjin.

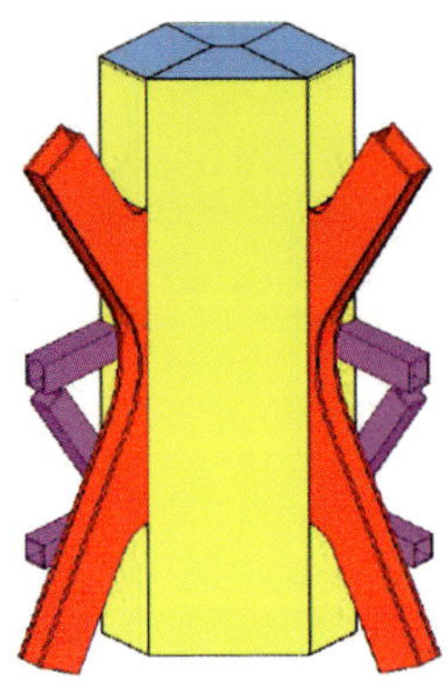

1b5
The multi-cellular mega column section also includes provision for connection to the belt truss
© Arup

Pushing the boundaries of conventional design

If Tianjin 117's engineers had limited themselves to solely using existing codes to design the structure, the tower would never have been built.

According to Chinese seismic codes, Tianjin Goldin Finance 117's height-to-width ratio should not exceed 8. But being both tall and slender, its ratio is 9.7 and, as such, required special approval from the National Expert Panel before its design could proceed to detailed design stage. Chinese design codes together with performance-based design principles and extensive linear spectrum-based and non-linear or time-history-based analyses were carried out for different levels of earthquake and wind loading combinations to justify the performance of the structure.

Arup also set new benchmarks for efficiency in the design of the tower's foundations by shaving 20m off the original 120m long piles. Extensive numerical analysis, together with fully supervised on-site trials, and review by internal and external experts supported Arup's case to reduce the pile size, against general rules in codes, to allow them to be more easily and safely constructed.

Arup also used Analysis and Design Engine (ADE) modelling software to quickly turn around the re-design of many structural components during the course of the project. Logic rules and constraints inputted by engineers createdthe parameters for modelling the structure as small details were changed, allowing the entire model to be updated reliably in a short space of time.

1b6
Tianjin Goldin Finance 117 Tower under construction
© Arup

1c
© Arup

Case study 1c

The Masterpiece (K11)
Hong Kong

CONTRIBUTORS: KIN-KEI KWAN + CHRIS CHEUNG

Client:	Urban Renewal Authority and New World Development Co. Ltd
Design architect:	Dennis Lau & Ng Chun Man Architects & Engineers (HK) Ltd
Arup's role:	Structural and geotechnical engineer
Year of completion:	2009
Height:	260m
Number of storeys:	64 (plus four basement levels)
Gross floor area:	100,000m^2
Use:	Hotel, residential and retail
Main contractor:	Hip Hing Construction

Rising up from the bustling streets of Tsim Sha Tsui on the tip of Hong Kong's Kowloon peninsula, is the 260m tall mixed-use Masterpiece (K11) Building. The Masterpiece is one of the tallest residential buildings in Hong Kong and sits on top of the K11 "art" mall – a unique retail experience, which integrates permanent displays of local artwork with shopping. The mall is named after the building's development code, assigned by the Urban Renewal Authority.

Tsim Sha Tsui is popular with tourists for its high end shopping streets, restaurants and museums. However, older buildings, which are no longer fit for purpose, blight parts of the district. Since 2001, the Urban Renewal Authority has encouraged regeneration of these plots of land and, with landowner New World Development Company, has led the way for the 68-storey Masterpiece to breathe new life into the area.

Tall and slender

The Masterpiece contains a hi-spec 21-storey hotel topped by 40 levels of apartments. Below, and extending beyond the tower's footprint, is an eight-storey podium structure which houses six levels of shopping of which two levels are in the basement and a further two levels of basement car park (Figure 1c1).

In plan, the tower is a chamfered rectangle measuring 80m wide by just 18m deep (24m maximum across its centre). The podium mainly extends beyond the tower's southern long-elevation.

With a height to width ratio of about 12, the building is classified as a very slender structure. The tower's internal arrangement of shear walls and its long, narrow floor plate perfectly suit residential occupancy and maximise net floor area, offering residents spectacular views across Kowloon and Victoria Harbour (Figure 1c2).

1c1
The mixed-use Masterpiece (K11): tower and podium
© Arup

1c2
Residents of the Masterpiece enjoy views across Kowloon and Victoria Harbour
© Arup

Form follows function

The Masterpiece's slim profile is one of its most obvious architectural qualities, but realising it required a great many structural issues to be resolved, resulting in its design being more structural engineer-led than architect-led. In fact, the building is an example of a "form follows function" style of architecture as its main structural features (a central core, which protrudes along the building profile, and belt trusses, which wrap around the tower in two locations) are expressed as part of the building's architecture.

Unusually also for a building of this height in Hong Kong, its main structural material is reinforced concrete – chosen for reasons of economy, despite being a heavier material than steel. The tower's main frame consists of a central core with coupled shear walls (Figure 1c3). Its podium comprises more conventional beam and column construction. Floor slabs are typically 150mm thick across the building.

When presented with this scheme, Arup's first challenge was to ensure that the building complied with deflection limits set in design codes for a slender building of this height. Under typical wind loading for this original structural system, the tower deflected more than was permitted (Figure 1c4).

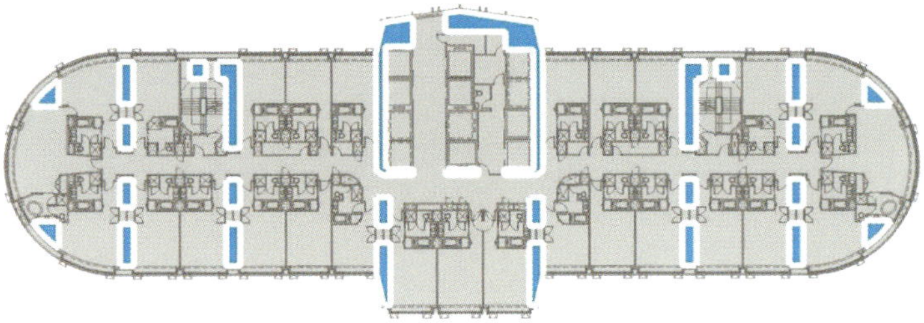

1c3
Original scheme: vertical elements on typical floor of the Masterpiece tower
© Arup

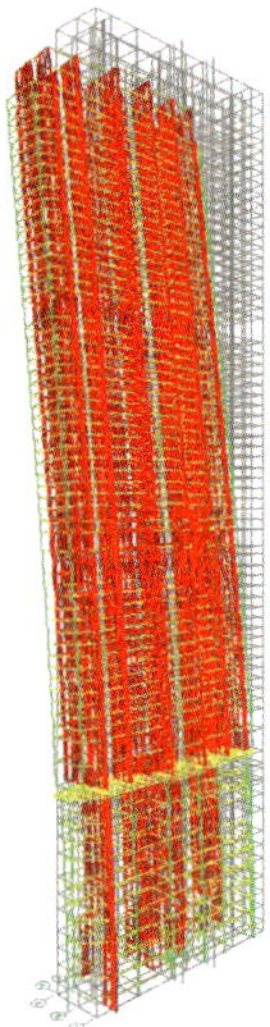

1c4
Deformed shape of the original scheme under wind loading (without outrigger)
© Arup

Belt (truss) and (outrigger) braces solution

To control deflection at the top of the tower, Arup designed a colossal outrigger system within the podium to stiffen up the structure (Figure 1c5). This involved adding a pair of outrigger beams (up to 10.5m deep by 3m wide with a 4m thick transfer plate) between levels seven and eight. These were connected to mega columns measuring 3.14m by 3.14m. The effect of stabilising the tower in this way also meant that the size of columns and walls in the tower could be reduced.

Embedded in the concrete mega columns are steel cruciform columns as well as heavy reinforcement, to enhance each member's stiffness. The combination of core, outrigger beam and mega column provides 65% of the building's wind load resistance. The remainder is provided mainly by shear walls in the tower.

Arup also rationalised the forces that the building would be subjected to under wind loading by carrying out wind tunnel tests on a scale model. This justified the use of a lower, more realistic wind loading value. It also meant that the building did not need a damping system to counter any sway, making the Masterpiece the most slender building in Hong Kong that does not employ a damping system.

While deflection of the tower under wind loading was now under control, introduction of the outrigger also had the effect of introducing torsion into the system. Moreover, with the building's shear and core walls orientated north–south, the only stiffness provided in the east–west (long) direction was via perimeter beams on each floor. Such a weakness in one direction would cause the building to become unstable and, without scope to add more walls to the building's interior, the solution had to involve more discreet strengthening.

Inclusion of two belt trusses between floors 25 and 27 and 47 and 48 aided the tower's lateral stiffness by sharing loads to more columns and remedied the problem. Due to the slenderness of the Masterpiece – and in contrast to more

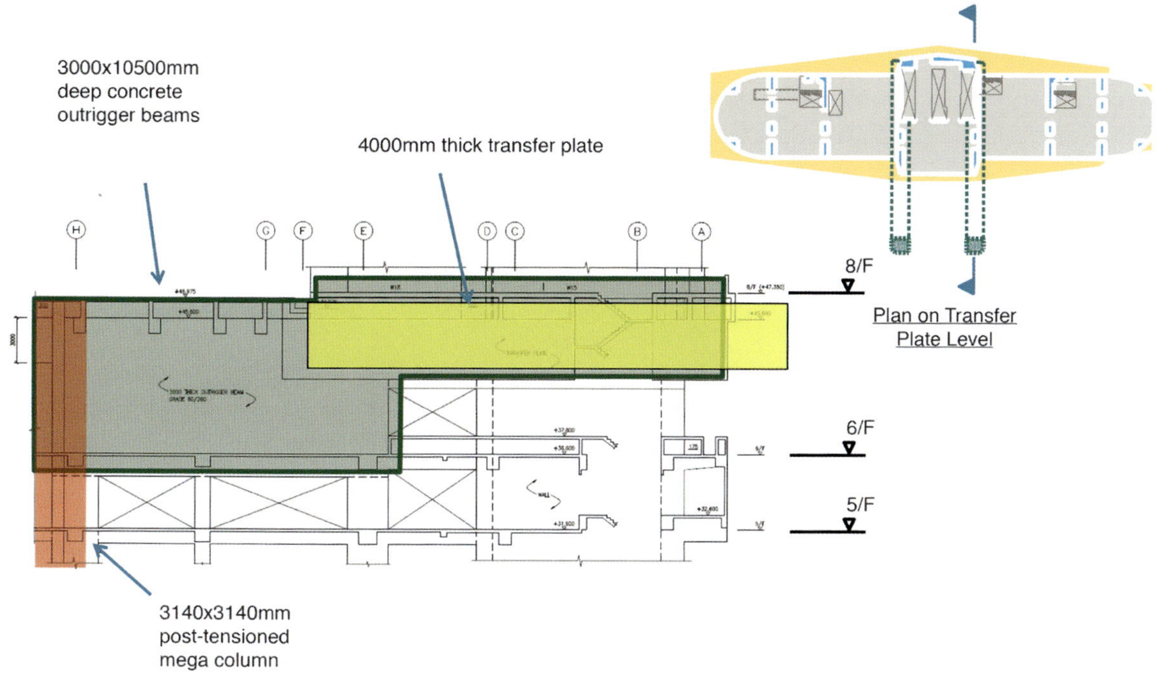

1c5
Cross-section through outrigger (left) and location of the outrigger in plan (top right)
© Arup

typically square high-rise buildings – the belt truss here is designed to increase minor axis stiffness. Again, to keep costs down, only diagonal members of the trusses are steel. Horizontal upper and lower chords are concrete and are integrated into the main building frame.

The masterstroke: vertical pre-stressing

1c6
Constructing the belt trusses
© Arup

The final piece in the jigsaw was to ensure that the mega columns and walls in the structure remained in compression under all loading combinations. Wind loads would induce tension in vertical elements of the building, potentially causing them to crack and elongate excessively. Pre-stressed tendons were introduced in key vertical members (Figure 1c7) to minimise tensile stresses. This results in a much stiffer building and less deflection under wind loads.

Adopting vertical pre-stressing – the second Hong Kong project for Arup – is the building's lasting legacy. Pre-stressing horizontally, usually in beams or floors to allow members to span longer distances without increasing their depth, is common practice, but pre-stressing vertically is relatively uncommon.

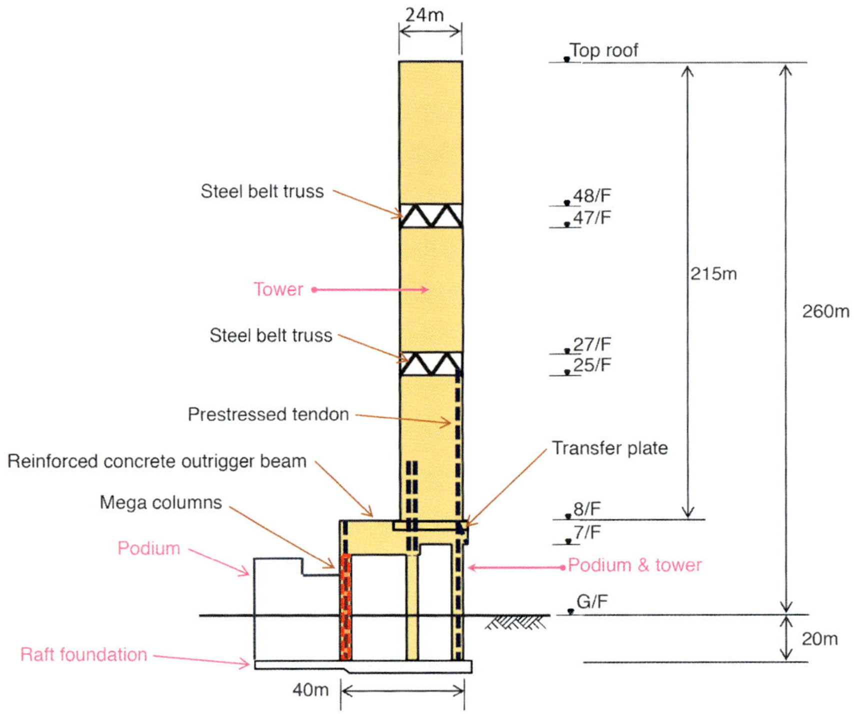

1c7
Masterpiece: the final structural system including mega column, outrigger, belt trusses and vertical pre-stressing
© Arup

1c8
Post-tension stressing undertaken in a recess at the top of a wall (left) and stressing at the soffit of the transfer plate (right)
© Arup

The success of this building largely comes down to the way that the client and project team members strived to accommodate these fundamental changes to floor plates and façades: the outrigger beam, for example, occupies two storeys in the podium, while the belt trusses create a very prominent architectural feature. Fortunately, the use of reinforced concrete, with its much shorter lead-in time than steel, allowed the many design changes to take place without delaying the overall construction programme.

While the Masterpiece is so named because of the artwork it contains, the term could equally be applied to the agility of its engineering design, ensuring that this building stands out from the crowd.

Basement construction

With the building founded on rock, construction of the Masterpiece involved excavating and supporting its 20m deep basement using a combination of diaphragm walls, propping and soldier pile walls. A 3m thick reinforced concrete raft supports the tower, while a 1.5m thick reinforced concrete slab with tension-resistant mini-piles supports the podium. Connections to adjacent subway tunnels at basement two level are accommodated between soldier piles.

1c9

Basement construction: diaphragm wall (left) and soldier pile wall (right)

© Arup

Chapter 2

Facilitating construction

Tall buildings with complex architectural forms or those positioned on areas of weak ground may never get built because of the risks they pose to construction safety and cost. Controlling these risks requires meticulous engineering design.

Through computer modelling, more predictable geometries for simpler fabrication and erection can be achieved. Engineers can also predict how these buildings will behave during construction and inform contractors where there is flexibility in the construction sequence and where there are certain limiting factors. The balancing act played out by certain structural members can even be modelled so that long-term behaviour is taken into account in their design.

Tall buildings all have to be well grounded, however difficult the ground conditions. This can lead to situations where the hidden engineering of foundations in the ground can be as impressive as the tower on top. In situ trials and testing, combined with a strict regime for checking the integrity of construction increases confidence and greater efficiencies in foundation solutions, vital in the most challenging soils.

Working closely with contractors and suppliers, Arup's designs are developed holistically to address all these complexities in foundation, façade and structure. Given the opportunity, they are the sorts of projects where the role of the engineer is most pertinent – to improve certainty – so that these schemes are economically feasible, safe to construct and the building industry can use the experience to move forward.

2a
© Zhou Ruogu Architecture Photography

Case study 2a

CCTV Headquarters
Beijing, China

CONTRIBUTOR: CHAS POPE

Client:	China Central Television
Design architect:	OMA Stedebouw BV
Associate architect:	East China Architecture & Design Institute
Arup's role:	Structural, geotechnical, security, fire, mechanical, electrical and public health engineer
Year of completion:	2012
Height:	234m
Number of storeys:	54
Gross floor area:	473,000m^2
Use:	Office
Main contractor:	China State Construction Engineering Corporation

The China Central Television (CCTV) Headquarters building in Beijing is one of the most unconventional structures in the world. The 234m tall building contains filming, production and administration facilities for the national broadcaster and is made up of four parts, which connect to form a single continuous "loop". These parts include two leaning towers of differing height, a 10-storey podium and a high-level "bridging" element, or Overhang, itself between nine and 13 storeys tall. Both the Overhang and podium connect the towers and are angular edifices with their noses pointing in opposite directions (Figure 2a1).

For Arup, there were already the usual tall building challenges of robustness and high seismic and wind loads to consider, but due to the CCTV building's unusual shape, added to this were unprecedented stress concentrations, untypical movement and, fundamentally, the question of whether such a building could be built safely and at an acceptable cost.

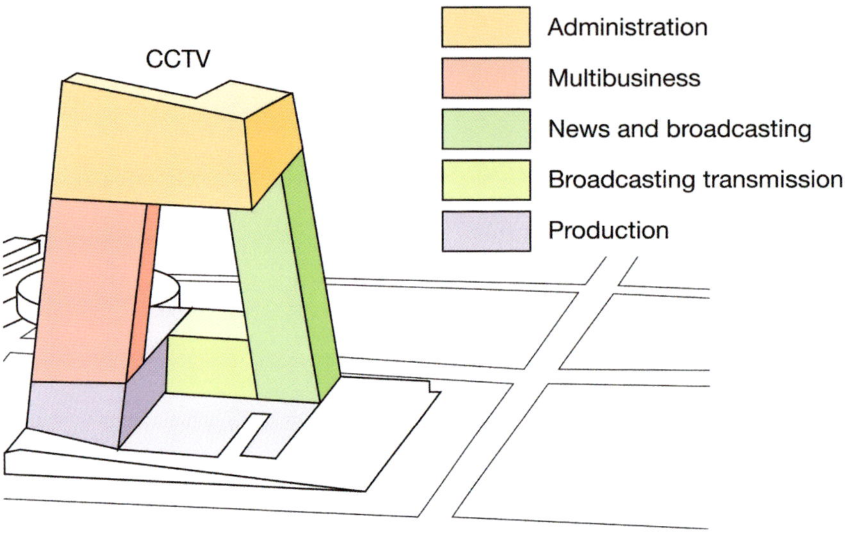

2a1
The four components of the CCTV building – two leaning towers, podium and Overhang – that each serve different functions
© Nigel Whale/Arup

2a2
The CCTV building's continuous "tube" structure, consisting of dense and sparse bracing optimised to accommodate different stresses in the building
© OMA

Continuous tube structure

Arup adopted a continuous tube structure for the building, where a "mesh" of steel or steel-reinforced concrete (SRC) columns, diagonal bracing and beams wrap around the building like a protective "skin". The team considered that the robust braced tube – with its multiple load paths – should be optimised to suit areas of high and low stress. In areas of high stress, the mesh would be dense and, in areas of low stress, it would be sparse (Figure 2a2).

Construction method influences design

Both the method and sequence of construction of the CCTV building influence the distribution of permanent loads in the building and the stresses that develop temporarily (during construction) and permanently (after completion). Therefore, right from the start, Arup had to consider a number of possible construction methods that would influence the building's design. This resulted in the consultant producing a special document, called the Particular Technical Specification (PTS), to communicate to other parties the various assumptions that had influenced the final design.

Specialist designers and contractors would need to understand, for example, the particular stresses and movements, which would be apparent during construction, particularly prior to connection of the Overhang and during and after its connection. These nuances would influence fabrication and construction.

Presetting to offset predicted movements

Construction of the CCTV Headquarters involved first building the steelwork to roof level in both towers. The podium was constructed concurrently. When the towers had reached their final height, cantilevering elements from the top of each tower were then incrementally erected (similar to the connection of two ends of a cable-stayed bridge) to create the Overhang (Figure 2a3).

Arup's structural analysis predicted that the Overhang would move 300mm downwards under its own weight at the moment when the cantilevering arms were connected. To accommodate this displacement, the PTS outlined that the main contractor had to preset the structure to ensure the building would be positionally correct on completion. The 300mm preset was achieved by the cumulative effect of every floor being constructed slightly out of position (Figure 2a4). In practical terms, this meant that the main contractor calculated every positional adjustment, which needed to be made according to its chosen construction sequence. For some members, these adjustments were made at the fabrication stage.

2a3
The Overhang was constructed by incrementally launching two cantilevers from the top of each tower
Images a-c © Frank P Palmer
Image d © Arup

a b c d

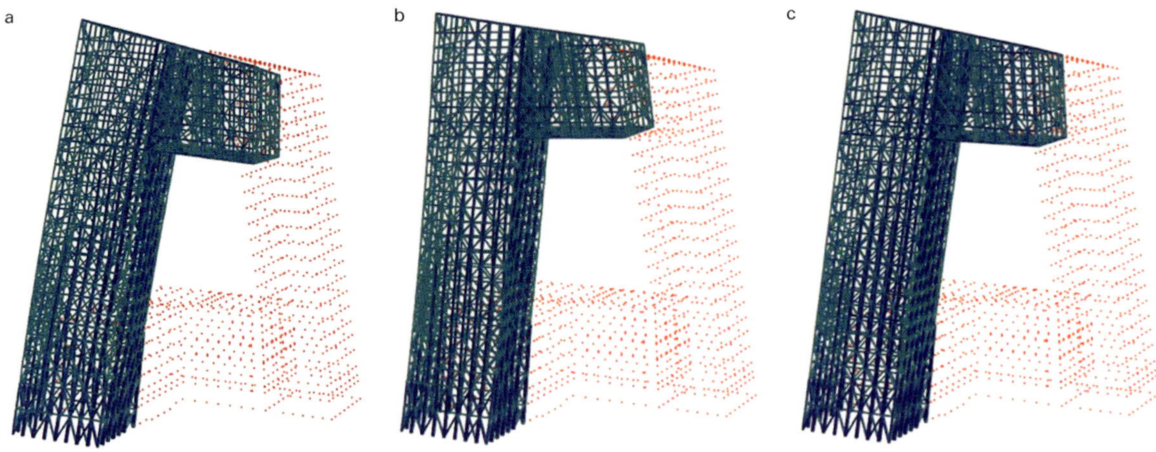

2a4
The philosophy of presetting the structure during construction
a: Calculations demonstrated that the tower would deflect under its own weight
b: Presetting the building upwards and backwards during construction offsets this deflection
c: There is "no deflection" under self-weight when the Overhang is connected
© Arup

As construction proceeded, the main contractor had to constantly monitor the movement of the building to verify that it was responding as predicted both by its own detailed "construction time-history" analysis and as specified in Arup's PTS. Any variations would then have to be re-analysed and adjustments made to subsequent floors.

Postponing installation of critical columns

The PTS also included other critical locations and stages where excessive movement or stress would have to be accommodated by manipulating the fabrication or erection process. In particular, Arup had calculated that the inside corner columns of the towers nearest the Overhangs would become highly stressed during construction. In structural engineering terms, these columns were designed as part of a system to resist seismic and wind loads rather than just vertical gravity loads. However, their location meant that they would inevitably end up taking the brunt of the weight of the Overhang. Since designing the columns to take this vertical load would have made them too large (and would have created a tendency to attract even more load, away from adjacent columns), a different solution needed to be found.

This came in the form of designing the tube steelwork to work under normal conditions without corner columns, so that adjacent columns and bracing would be beefed up to resist the weight of the Overhang. Then, several months after the structure had been completed and forces had settled in the fully connected "loop", the columns could be safely installed without forming part of the gravity load path to the ground. The nature of the continuous tube structure

2a5
The four inside corner columns beneath the Overhang (highlighted in red) were installed after the rest of the building's steel frame had been completed
© Arup

is such that, day-to-day, many members are underused, which leave spare capacity in the structure to resist greater forces in the event of a severe earthquake, storm or other extreme event.

As an indication of the extreme stresses that parts of the structure were designed to resist, the steel sheeting which supports the concrete slabs uses 15mm thick plate on key floors instead of thin 1mm thick proprietary decking as typically used for office buildings.

Loading scenario analyses

Arup considered two scenarios to analyse the effect of adding varying amounts of "dead load" (the weight due to permanent structural members, fixtures and finishes) before and after the Overhang was connected. An "upper" bound analysis was developed for the scenario where more additional dead load was added while the Overhang was still under construction, putting the largest stresses in the towers in the temporary condition. A "lower" bound analysis was also carried out for the scenario where a larger proportion of this dead load was only added to the already connected Overhang. This scenario would result in higher stresses being developed in the Overhang (which acts as a prop between the towers after connection).

The outcome of the upper and lower bound analyses meant that structural members were designed for a worst load-case scenario, which gave the main contractor flexibility to apply dead loads according to its own construction sequence rather than one prescribed by Arup. In fact, for the building façade to be ready for the 2008 Beijing Olympic Games, façade erection had to start early and before the towers were connected, despite its weight inducing higher stresses in the tower. The façade was designed and installed to accommodate the expected movements and stresses of the building during construction, as a result, and the tower steelwork was designed for the effect of additional façade loads prior to Overhang connection.

Connecting the Overhang

Undoubtedly the most tantalising part of the CCTV build involved constructing the Overhang. This was where 75m of cantilevering steelwork stretched out from the top of each tower, up to 162m in the air until the moment each great arm made contact with the other, uniting the two towers. But imagine also the risks: high winds causing the unsupported ends of the cantilevers to move, large temperature variations causing the steelwork to expand and contract, and some of the highest possible stresses developing in the tower.

Arup outlined in the PTS the exact conditions under which connection should take place. It stipulated that this would be when the two towers were at a uniform temperature, and movements across the joint were at a minimum. This translated to connection only being possible just before dawn on a windless day. The connection was also required to have the capacity to take on high stresses at that location instantaneously in order for the forces to propagate quickly through the towers. The PTS recommended installing a series of strong permanent linking members which could be installed in a way which initially allowed movement, but could quickly be "locked off" to create a stiff connection.

Arup also specified that the building's global and relative movements should be monitored for seven days prior to connection (Figure 2a6). The actual measurements recorded daily relative movements of up to 10mm between the towers.

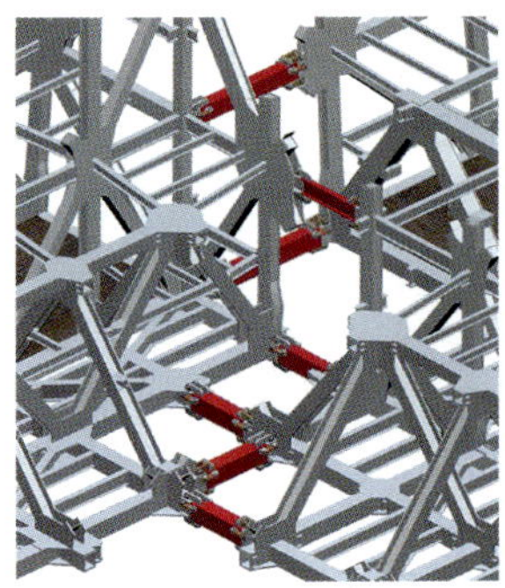

2a6
Structural model showing connection members (in red) for Overhang
© China State Construction Engineering Corporation

On the day before "connection day", the contractor lifted all seven connection elements into position, noting just 1mm of tolerance between the two towers. The following morning, with favourable weather conditions and in the space of just a few minutes, the bolts were simply tightened and the Overhang was connected (Figure 2a7).

Despite the complexities of the CCTV building's design and construction, there were no major delays on-site, nor any unexpected building movements.

The CCTV building's gravity-defying architecture was made possible by an enlightened design and construction team working closely together. Arup's PTS was key in communicating the assumptions on which the design was based and the outcomes required, so that the entire project team was able to understand the nuances of the build. Inclusion of a PTS has since become a more common feature of complex building design in China.

a

b

2a7
"Connection day": the cantilevering arms of the Overhang prior to connection (a), and after connection (b)
© Arup

2a8
Consideration of the construction process throughout the CCTV building's design ensured its success
© Arup

2b
© Atkins

Case study 2b

Vincom Landmark 81
Ho Chi Minh City, Vietnam

CONTRIBUTORS: KIEN HOANG, THE TRUONG + HUONG PHAN

Client:	Vingroup
Design architect:	Atkins (Hong Kong)
Associate architect:	VNCC (Viet Nam Construction Consultants)
Foundation contractor:	Bachy Soletanche Vietnam Co. Ltd.
Arup's role:	Structural, geotechnical and wind engineer up to construction drawings stage
Year of completion:	2019
Height:	461.2m
Number of storeys:	81
Gross floor area:	142,000m^2
Use:	Residential (L6–L45) and hotel (L46–L63)
Main contractor:	Coteccons

The fertile banks of the Saigon River in Ho Chi Minh City, Vietnam are cultivating a new kind of crop – tall buildings – to create a fresh and dynamic urban landscape. The tallest currently under construction is Landmark 81, a tower that resembles a bundle of bamboo canes to reference the area's agricultural past and a nation's much revered home-grown resource. With a hotel at the top, apartments occupying lower levels and a five-storey retail podium, the building forms the centrepiece of a new residential development.

Most recently used as a dockyard, the site's ground conditions, as for most of Ho Chi Minh City, are poor for construction. The top 30m of muddy ground is virtually unusable in structural terms and, combined with the heavy weight of a 461m tall reinforced concrete building, posed significant foundation challenges for engineers. Below ground level, Landmark 81 also has a 12m deep, three-storey basement. Allowing for local contractor experience, large diameter bored piles or rectangular barrette piles up to 90m long emerged as the most viable options.

Flexible and fast-paced

Taking the scheme to even greater extremes, the client's fast-track programme meant that the foundation solution had to be developed at the same time as the superstructure, which meant that it would need to have some degree of flexibility to accommodate changes. Arup's strategy was to create a broad-based foundation design that could be refined as more detailed information became available.

Past experience offered a conservative starting point for vertical building loads, while local design codes provided typical wind and seismic loads for the region. The assumption was that following optimisation of the design, final vertical load values would be smaller than originally assumed. In the case of wind loading, wind tunnel tests would provide a more accurate picture, as wind was expected to provide a smaller load than that resulting from design code.

The rectangular barrette pile solution became the front runner early on due to its greater capability for accommodating different loads by varying its size and orientation, compared with circular bored piles. A preliminary scheme was developed while site investigation and wind tunnel tests were still being carried out and superstructure design continued.

There then remained the question of whether the barrette piles should be shaft-grouted to increase their shaft friction and, hence, load carrying capacity. The procedure involves injecting high-pressure grout around the pile's shaft to increase the friction acting between the pile and the ground. Carrying out an on-site trial would indicate the effectiveness of shaft-grouting to engineers. It was considered that, for the highest load capacities to develop in a pile, they would also need to be constructed to a very high standard.

2b1
Wind tunnel test for Landmark 81 Tower
© Rowan Williams Davies and Irwin Inc.

However, it became clear around this time that there were very few local contractors that could take on such a large job or had workers experienced in following stringent guidelines for pile construction. Two contractors with an international background were finally shortlisted for selection in this project. The project would mark a change in mindset for the local piling contracting community.

To grout or not to grout?

Two 1m wide by 2.8m long test piles were constructed on site to depths of 85m and 80m below ground. The shorter barrette pile was shaft-grouted.

Load tests were carried out on the piles to verify design assumptions. An Osterberg cell was used to generate loads and confirm that parameters adopted for design, together with the axial shortening of the piles, were within an acceptable range. Further tests verified that the piles were constructed to within acceptable vertical and geometrical tolerances.

The test piles demonstrated that greater load capacities than expected from conservative design calculations could be achieved, but relied on a high quality control on the shaft-grouting procedure. Since it would take between two and three days to install each pile, reducing their number would ensure that the project stayed on programme.

2b2
Osterberg cell installation in the reinforcement cages of a barrette pile
© Courtesy of Bachy Soletanche (2015)

Table 2b1 Results of the two test piles where only one is shaft-grouted

	Test Pile 1 (non-shaft-grouted)	Test Pile 2 (shaft-grouted in segment)
Size (mm)	1,000 × 2,800	1,000 × 2,800
Founding level (mPD)	–85.0	–80.0
Design load (kN)	31.300	35.800
Maximum test load (kN)	78.250	89.500
O-cell load (upward and downward) (kN)	31.125	44.750
Downward movement of the O-cell base (mm)	41.5	17.1
Upward movement of the O-cell top (mm)	16.3	18.3

Table 2b2 Skin friction and end bearing values. The base resistance is not fully mobilised in TP2 as the test was terminated before a sufficient amount of settlement could occur

Soil Profile	Top Level (mPD)	Trial Pile 1		Trial Pile 2 (shaft-grouted in segment)	
Filling soil	-1.7 to -3.2				
Organic silty CLAY	-28.3 to -29.0	Unit Skin Friction	50 kPa	Unit Skin Friction	50kPa
Sandy CLAY, firm	-30.5 to -35.5	Unit Skin Friction	140 kPa	Unit Skin Friction	200 kPa (SG)
Clayey SAND, dense, fine to medium	-63.8 to -64.6	Unit Skin Friction	140 kPa	Unit Skin Friction	260 kPa (SG)
Dense, fine to medium SAND	-71.0 to -87.5	Unit Skin Friction	170 kPa	Unit Skin Friction	270 kPa (SG)
				End Bearing	2,500 kPa
Very dense, coarse SAND	-92.0 to -96.7	Unit Skin Friction	150 kPa	**Remarks:** Not to scale; SG = shaft-grouted section	
		End Bearing	4,507 kPa		

2b3

Installation of reinforcement cage during construction of the barrette pile
© Courtesy of Bachy Soletanche (2015)

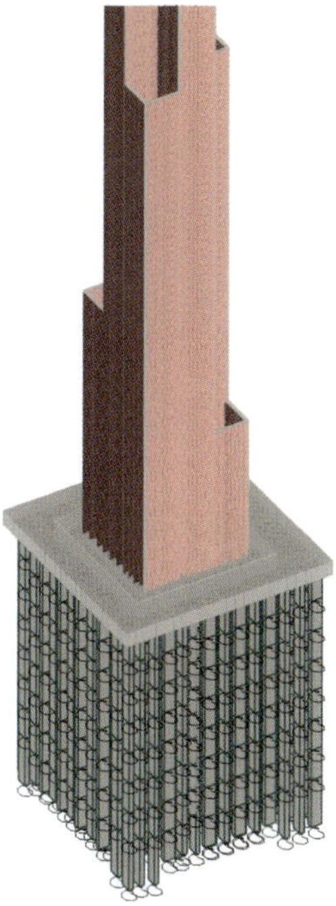

2b4

Computer model showing interaction of the ground, raft and pile for tower loads
© Arup

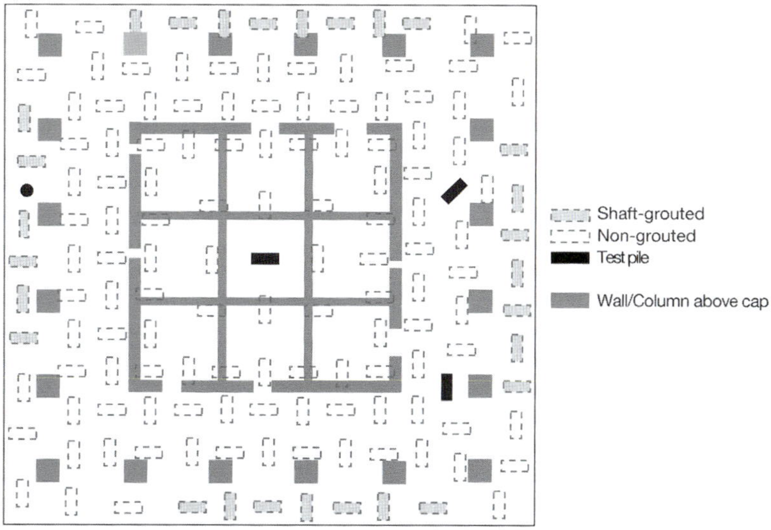

2b5

The final pile layout shows barrette piles orientated to suit the specific loading requirements of each location
© Arup

Modelling and refining

For a scheme of this scale and complexity, modelling and analysing the foundation solution by computer provided the most realistic and efficient means of understanding and refining the design. The initial foundation study considered a range of thicknesses for the basement (B3) raft slab between 3m and 10m, with 1m × 2.8m barrette piles positioned at least 2m apart.

The most up-to-date vertical and lateral loadings were then applied to a computer model of the design to determine the stress distribution in the piles. An iterative process then followed where pile sizes, orientations and positions were varied until the capacity of every pile was satisfactory for each thickness of raft investigated. The process highlighted which piles would need shaft-grouting to increase their capacities.

Engineers weighed up the benefit of a thicker raft spreading loads wider, with the time and cost involved with excavating deeper, before coming up with the final raft profile. The raft is, in fact, 8m thick at its maximum, reducing to 4m at its minimum, making it the thickest pile cap constructed in Vietnam to date.

When the results from the pile trial, wind tunnel tests and site investigation, as well as the actual superstructure loading (which was reduced compared to the preliminary design) and construction cost of grouted and plain piles became known, the information was used to refine the pile design. The final scheme developed to include 147 barrette piles, of which 28 piles were shaft-grouted. Piles are 1.2 × 2.8m and 1.0 × 2.8m Grade C35/40 and C40/50 concrete, extending 69m to 74m below basement respectively.

Because of the great stiffness of the superstructure and the pile layout arrangement, gravity loads spread quite evenly among the piles; whereas the perimeter piles are governed by seismic load and wind load cases due to the push–pull effect.

Unprecedented design and build

The barrette piles were constructed from a piling mat at ground level using methods common to diaphragm wall construction. Some piles were designed and installed with columns cast on top to form part of a temporary support system for the long spanning sections of the ground floor.

Shaft-grouting was carried out using *tube a manchettes*. These are tubes that allow high-pressure grout to pass into the ground around the pile and are installed at the same time as pile reinforcement. All piles also underwent extensive monitoring and testing, exceeding local code requirements to verify the build quality.

It took just two and a half months for the piles to be constructed, the result of a design that placed local skills and capability at its heart. The project owes its success to the time invested in on-site testing to verify the piling solution together with developing a rigorous installation method to ensure a high-quality build. It proves that poor ground conditions are no barrier to building tall.

Basement: build it faster

Both top-down and bottom-up construction techniques were employed in the basement for programme gains. Top-down construction is generally faster than bottom-up since floors above can be constructed simultaneously with the basement, while at the same time minimising propping. However, it involves a more complex build sequence. Bottom-up construction was considered more appropriate within the extent of the tower, due to the greater loading demand of the tower columns, which would make the top-down methodology less viable.

The basement uses an inner and outer ring of 800mm thick, 32m deep diaphragm walls to create a "donut" shaped top-down excavation zone, leaving the centre of the "donut" (the area of the tower) for bottom-up construction.

The top-down construction sequence involved casting the ground floor slab and leaving muck openings for excavation below. The B1 slab was cast when excavation had reached the appropriate level, again with muck openings left through which the lower floor could be excavated. B2 and B3 levels were constructed using the same method.

The B3 raft beneath the tower was cast in two layers to reduce the volume of concrete in each pour, which would allow better control of its temperature and limit cracking. Pipes with iced water were incorporated in the raft to help maintain temperature within a predetermined acceptable range.

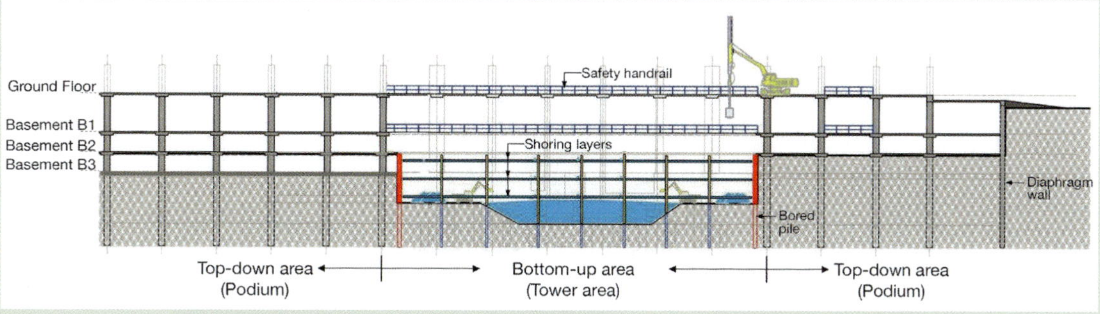

2b6

The three-storey basement was built in the most time-efficient way, using top-down construction methods where possible
© Arup

2c
© P&T/Goldin Properties

Case study 2c

Tianjin Goldin Finance 117 Tower
Tianjin, China

CONTRIBUTOR: GARY GE

Client:	Tianjin Hi-Tech New Star Property Development Co Ltd
Design architect:	Palmer & Turner Consultants (Shanghai) Ltd
Associate architect:	East China Architectural Design & Research Institute Co Ltd
Arup's role:	Structural, geotechnical and wind engineer, façade consultant
Year of completion:	2018
Height:	597m
Number of storeys:	117
Gross floor area:	370,000m^2
Use:	Office and hotel
Main contractor:	China State Construction Engineering Corporation (CSCEC) Third Bureau

The story of what lies beneath the 597m tall Tianjin Goldin Finance 117 Tower is as impressive as that of the building above. Indeed, had the foundation challenges of this colossal tower not been solved, it might never have been built.

Tianjin is a city close to the sea, where its fine soils are not conducive to supporting heavy loads – unless a great deal of engineering mastery is brought into play. Put simply, the engineering challenge here was to support the weight of an exceptionally tall, slender tower on soft soils in a seismic area. With no precedent for any similar kind of building subjected to the same constraints in the region, Arup embarked on a journey into the unknown to find a solution.

Soft ground challenges

The site of the tower was once occupied by a fishpond, which had been backfilled with rubbish and waste concrete to a depth of 3.7m below ground level. Below this, the ground (identified by boreholes as part of a previous site investigation) had only two relatively stiff strata – one at a depth of 100m below ground level and the other at 120m. Piles 1m in diameter and 120m long were calculated to be the most reliable option, making them some of the longest in China.

Bored piles are the most common foundation type for tall buildings in Tianjin and take the form of reinforced concrete columns in the ground. They control settlement and transfer vertical loads coming down the walls and columns of the building to the soils. They develop their capacity by mobilising the frictional resistance of the outer perimeter of the pile shaft against the soils. The end of the pile, which bears onto the ground, also contributes a little to the pile's load carrying capacity. At Tianjin, the soft soil characteristics of the ground meant that each pile's shaft friction would have a much greater role to play in supporting the building than its end-bearing capacity.

As such, the foundation solution needed to maximise pile shaft surface area to achieve the capacities required. This meant a large number of long, slender piles would be needed.

Since bored piles had only been installed to depths of around 75m, at most, in the region, there were still concerns over whether such long pile lengths could be constructed vertically and without suffering collapse. If the piles were raked, for example, they would exert additional load on other piles, potentially causing overstress on some piles.

Other foundation options

Other foundation options were considered, but each had their own limitations. Driven steel tubular piles were deemed too costly at more than double the price of bored piles, and might also suffer corrosion from groundwater. Barrette piles, similar to deep diaphragm wall panels, were also considered. The benefit of barrette piles over bored piles is that given the same concrete volume, greater surface area can be provided for shaft friction development. However, local contractors lacked experience in constructing them.

Arup opted for bored piles (Table 2c1) using high strength C50 tremie concrete (cubic compressive strength of 50N/mm^2 with a reduction factor 0.7 to 0.8 applied in the design for piles cast under water) with the view to optimising the solution through on-site trials.

Table 2c1 Foundation option study

Issue	Bored Pile	Driven Steel Tubular Pile	Barrette
Construction Method	Commonly used	There were case histories with comparable height	Seldom used, may have difficulties finding competent contractors
Settlement (from past experience)	Usually within predictable range	Often greater than predictions	No actual data
Approval	More straightforward	Straightforward	Envisaged difficulties due to lack of past experience and competent contractors
Construction from Ground Level	Yes	Yes	Yes
Impact on Adjacent Constructed Piles	Less effect	Whilst the frictional resistance of adjacent piles can be enhanced a little, negative squeezing effect could not be avoided	Less effect
Material	High strength concrete may be used, but little precedence	Available	High strength concrete may be used, but little precedence
Cost	1X	2.3X	1.3X
Overall	Recommended	Not recommended	Not recommended

Pile trials

The trials consisted of eight piles located in two different areas near the tower footprint and with similar ground conditions. Four piles 1m in diameter and either 100m or 120m long were tested in the first trial group. Each pile was constructed at original ground level, with double sleeves over its uppermost 25m to remove the effects of friction as far as possible and reflect the depth of basement. All piles were also subjected to shaft and toe grouting to determine their effects on the pile capacity. The piles were loaded to 42MN which was equivalent to twice the calculated ultimate load capacity of the 100m long pile and 1.5 times that of the 120m long pile. The intention of maximum test load being higher than the calculated ultimate capacity aimed to justify a set of more representative design parameters than interpreting conservatively from the site investigation results.

The first trial group demonstrated that 100m long bored pile with shaft and toe grouting would be adequate to support the working load with a safety factor greater than 2. The second trial group of four piles aimed to verify the contribution of pile capacity from shaft grouting and to further justify the characteristic value of single pile capacity of working piles. As such, all piles were subjected to toe grouting in the second trial but only two were shaft-grouted.

Table 2c2 Results of the two pile trials (All piles adopted shaft grouting and toe grouting, except S3 and S4 where only toe grouting was adopted)

Trial 1

Pile No.	Dimension	Construction length (m)	Effective length (m)	Proposed final test load (kN)	Vertical disp. (mm)	Ultimate capacity (kN)
3	Φ1.00×120m	120.57	95.50	42,000	55.72	≥42,000
6	Φ1.00×120m	120.51	95.50	42,000	58.92	≥42,000
9	Φ1.00×100m	100.84	76.00	42,000	50.54	≥42,000
12	Φ1.00×100m	101.13	76.00	42,000	52.25	≥42,000

Trial 2

Pile No.	Dimension	Construction length (m)	Effective length (m)	Proposed final test load (kN)	Vertical disp. (mm)	Ultimate capacity (kN)
S1	Φ1.00×100m	98.49	76.00	42,000	47.62	≥42,000
S2	Φ1.00×100m	98.40	76.00	42,000	52.54	≥42,000
S3*	Φ1.00×100m	98.33	76.00	42,000	56.96	≥42,000
S4*	Φ1.00×100m	98.27	76.00	42,000	93.54	37,500

The trials proved that the 100m long piles with a diameter of 1m with shaft and toe grouting would be sufficient to support the maximum pile load of 16,500kN with a safety factor of more than 2. The results from the pile trials also helped estimate the envisaged settlements of the pile group and check whether they would be within acceptable limits.

Tackling construction issues

Pile construction involves first boring into the ground using a drilling rig where bentonite slurry is pumped into the hole to support its sides and prevent collapse. When the hole has reached its required depth, a cage of steel reinforcement is lowered in and concrete is poured into the hole, displacing the bentonite slurry. In the final stage, each pile's toe and shaft are grouted to enhance its capacity (shaft grouting is exempted for those piles subject to low loads).

Close supervision and high-quality workmanship played a prominent role in yielding the highest possible pile load capacities for the project. Arup engineers took on a full-time site inspection role to identify and smooth out construction-related issues which would affect the pile quality.

The whole process of preparing the steel reinforcement cages and ensuring each 30m section was perfectly coupled with the next, through to checking the bentonite compositions and renewing it at the right time, was closely supervised. Concrete samples were also examined to check that the required C50 tremie concrete strength was being achieved. Extra checks on the instrumentation and that the cables were adequately attached to pile cages were carried out to ensure that the results from multiple data loggers and sensitive instrumentation would be accurate and could be cross-analysed and verified.

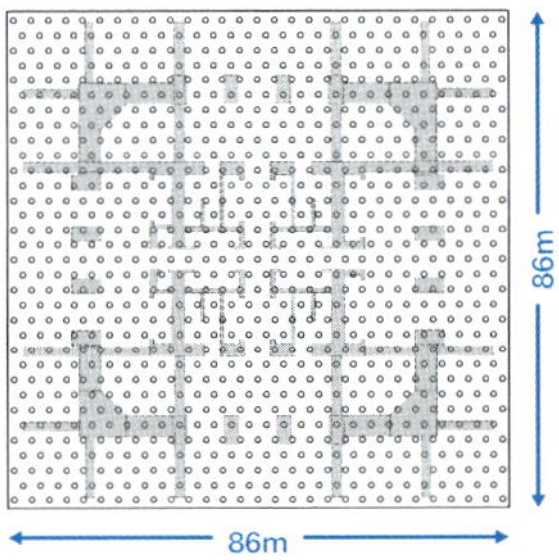

2c1
Final pile layout
© Arup

2c2
Loading up a pile to 42MN during the trial
© Arup

2c3
Pile drillhole construction
© Arup

2c4
Bentonite slurry preparation and treatment
© Arup

Shorter piles lead to cost savings

In the final pile layout, a total of 941 piles underlying an 86m long by 86m wide, 6.5m thick pile cap were constructed. Detailed design of each pile had to take into account different loading combinations, including seismic and wind loads, as well as gravity loads and uplift to determine the worst case scenario for each pile. To ensure the efficiency of the scheme, only a small portion of working piles (100m long and 1m in diameter) required a high proportion of reinforcement on top of using C50 tremie concrete. Other piles were designed to specific reinforcement levels based on the actual pile load from the design. Toe and shaft grouting was adopted for a majority of the piles, whilst toe grouting only was used for a small proportion of the piles.

The trials validated the use of piles 20m shorter than originally calculated, which had a significant effect on construction time and cost – about 30% cost savings per pile could be achieved. But perhaps more importantly, the trials and detailed foundation design gave the wider engineering community, especially local designers and contractors, the confidence that long piles in weak ground could be built successfully.

2c5
Installation of steel reinforcement
© Arup

2c6
Enhanced site inspection and working methods improved the quality of construction including coupling reinforcements (right) after certain out-of-tolerances were spotted during trials (left)
© Arup

2d
© New World China Land Ltd

Case study 2d

Tianjin Chow Tai Fook Finance Center
Tianjin, China

CONTRIBUTORS: ANDY LEE + WING CHIU

Client:	New World China Land Ltd
Design architect:	Skidmore Owings & Merrill (SOM)
Executive architect:	Ronald Lu & Partners (RLP); ECADI
Arup's role:	Façade and building maintenance unit consultant
Year of completion:	2019
Height:	530m
Number of storeys:	97
Gross floor area:	252,144m^2
Primary use:	Mixed – office, retail, 5-star hotel and serviced apartments
Main contractor:	China Construction Eighth Engineering Division

The Tianjin Chow Tai Fook (CTF) Finance Center is a landmark building located in Tianjin's new central business district, close to Bohai Bay. At 530m tall, the tower has a commanding presence, but it is the sinuous curves of its four elevations that make it so eye-catching. Made up of thousands of glass panels, the building's façade resembles the skin of an amphibious creature with shimmering scales, fresh out of the water.

As the building tapers with height, its floorplate morphs from square to circle, while convex and concave curves glide up the elevations to create an elegant and streamlined form. Initially, it was hoped that curved panels could be fabricated to follow the intended façade profile. But curtain wall manufacturers in China had limited experience of cold-bending curtain wall units and there was concern that there could be significant risk in terms of time, cost and future maintenance to include them on a project of this scale. Flat panels were the preferred practical option; however, this would raise the issue on how the curved building profile could be achieved without compromising the authenticity of the architect's design.

Starting from formidable complexity

With each floorplate being a different shape, storey-height curtain wall panels would need to be manipulated in some way to marry up with the floors above and below, since they could not be bent. Using complex equations to define the curves of each floor, and using a combination of tilting and offsetting the panels within their frames, a faceted curtain wall system was proposed.

The majority of the curtain wall panels would be asymmetrical units, each with four different inner lengths and four different corner angles. In total, 24,910 unique types of insulated glass unit (IGUs) for the vision and spandrel panels would be required. The panels were intended to be installed at variable angles along the entire tower façade, and would be tilted and/or offset within their frames in order to suit the curved building profile. Due to the undulating form of the façade, the degree to which the panels would be tilted and offset would have varied from floor to floor.

Paramount to the success of the building was for these panels to be fitted accurately. But with so many non-standard panels the potential for error would be high, as each panel's geometry would need to be set out individually for initial fixing, then checked and monitored during installation. Minute deviations in positioning lower down the building would be magnified further up and jeopardise the entire built form.

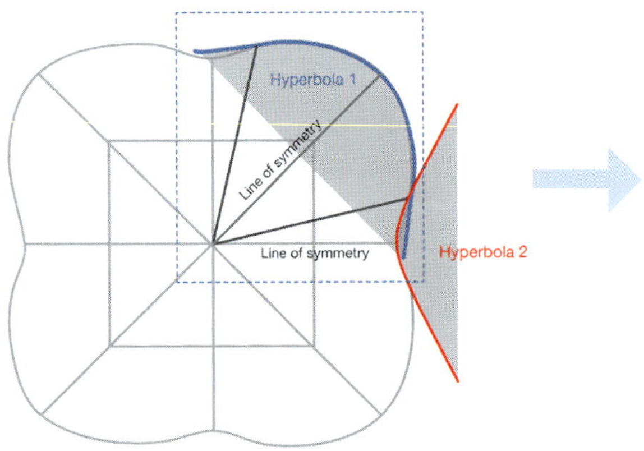

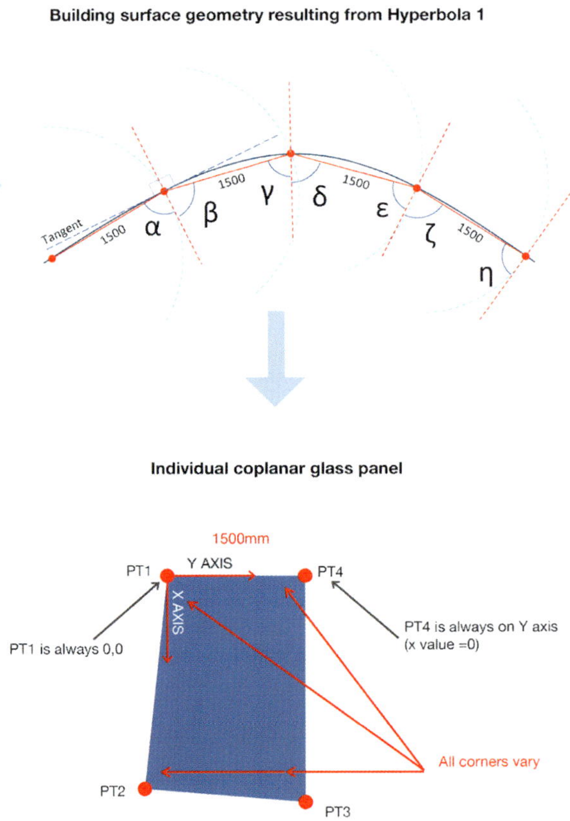

2d1
The original curtain wall panel design would be more complicated to fabricate and install due to variable corner angles (α, β, ϒ, δ, ε etc.) at different levels (top right) and variable side lengths of the asymmetrical glass panels (bottom right)
© Adapted from RLP/GT

Concern also emerged over the fact that if every panel was angled or offset differently to suit the curved building profile, the impression might be more random than the architect had originally intended.

Fabrication would be tricky and time consuming too, since the panels had different corner angles and side lengths. Curtain wall manufacturers also indicated that they might need to fabricate as much as 50% more panels on an order of non-orthogonal panels due to learning errors early on in the manufacturing process. So simplifying the panel geometry and reducing the number of panel types would reduce the likelihood of error, thus time and cost, during fabrication.

Working with Gehry Technologies (GT), Arup's optimisation study simplified fabrication and installation while adhering to the original design as far as possible.

Simplifying the geometry: using circles instead of equations to set out the panels

The first task was to optimise each panel's geometry by defining the curves of the building in a simplified and logical manner. This work involved rationalising the hyperbolic and elliptical equations which defined the architect's original curves, and redefining them using arcs of a circle of fixed radius. Working closely with the client and architect, another exercise to minimise the variation in geometry was carried out and resulted in the difference between this and the original geometry being an increase in total floor area of 68m^2.

Arup also proposed to split each floorplate into four segments, so that each quarter could then be split in half to create a mirror image of itself to reduce the number of different panels per floor (Figure 2d3). This would have a significant effect on speeding up fabrication and reducing complexity for each panel, as equipment would only need to be set up once for each different angle and radius. Without these measures, the Tianjin CTF Center's panels could have taken up to six times longer to fabricate due to the time taken to set up a computer numerical control (CNC) machine.

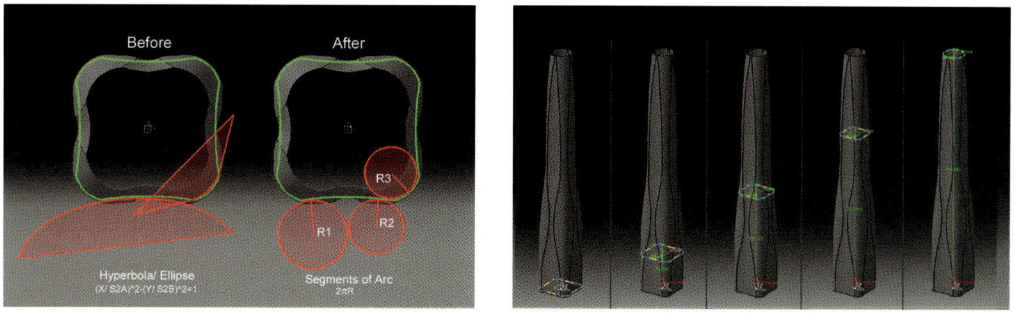

2d2

The curves of each floorplate were originally defined by equations for ellipses and hyperbolas, which were then rationalised into the arcs of a circle of fixed radius (left); the variation in surface geometry was therefore minimised to simplify the tower shape (right)

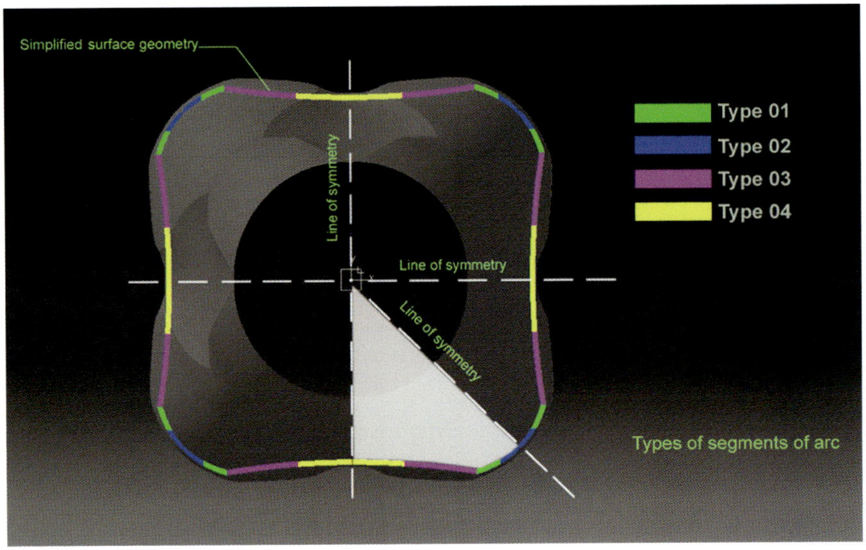

2d3
The simplified surface geometry showing lines of symmetry to reduce the number of panel types
© RLP/GT

Optimising the tilt angle and offset for each panel

Using parametric modelling, the range of corner offsets required for each panel was optimised. The analysis revealed that the maximum gap that would need to be accommodated was 62mm rather than 70mm and that this could be achieved using 20 rather than 24 different sub-frame designs. The range of angles that needed to be accommodated was optimised, and the number of different types of waterproofing joints required between panels was also reduced.

A specially designed adaptor in the mullions helps to accommodate different angles for each panel and slight variation in panel joint width (+/– 2.5mm) (Figure 2d4).

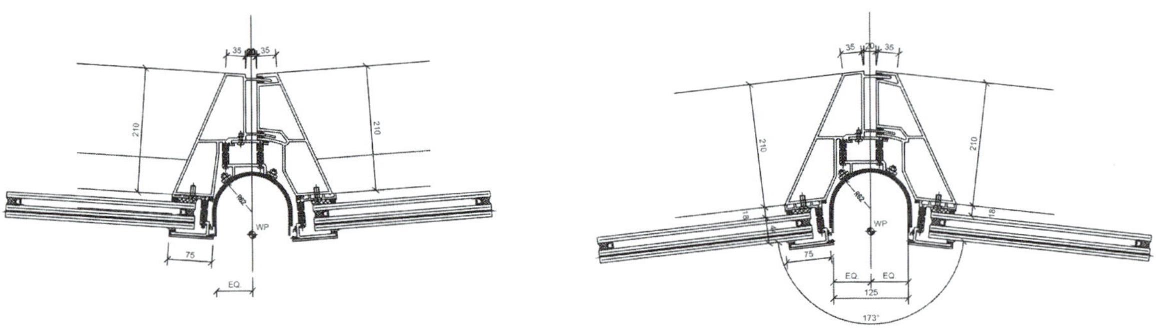

2d4
Mullion design includes a variable-sized pocket to accommodate different tilts or offsets
© Arup

The result: realising the intent with reduced number of panels

By rationalising the geometry of the panel to be symmetrical and into either 1,500mm wide rectangles or isosceles trapeziums, with the opposite side length deviating only slightly from 1,500mm, the unique types of IGU could be reduced from 24,910 to 1,308. Further input from the main contractor reduced this number to 1,219.

Where full height curtain wall panels were required, instead of separate vision and spandrel panels, numbers reduced from 730 to 339.

In all, four months were spent optimising the façade, but the effect was a two- or three-fold reduction in installation time and immense reduction in overall risk to the project.

The only question that remained was whether the optimised tower profile would still have the same impact as that intended by the architect. Statistically, 87% of the building envelope is within 20mm of the original design. And the optimised façade would lead to a more consistent glazing pattern and reflection as adjacent panels grouped by the same arc radius would be identical. Inevitably, there were still some locations where panels would need to be offset from the original arc line to follow the architect's intended building curve; however, these panels would be evenly distributed and fewer in number than originally envisaged.

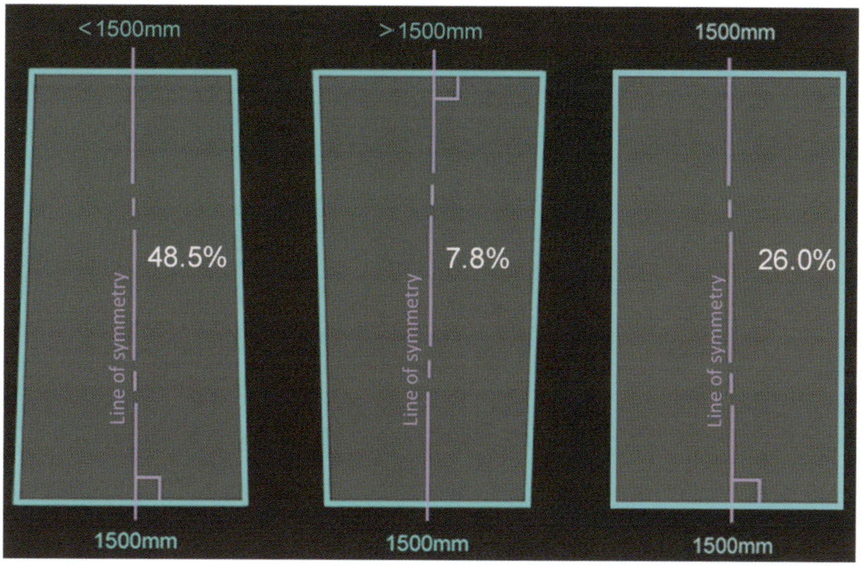

2d5
Optimising the panel size into rectangles or isosceles trapeziums (percentages show the proportion of each panel type)
© Adapted from RLP/GT

Chapter 3

All in one, vertically integrated

The viability of building tall rests not just on how to physically stack floors on top of one another, but on creating the mechanical, electrical and plumbing (MEP) systems which allow these developments to function efficiently.

Tall buildings typically devote whole floors to MEP at regular intervals to overcome the problems of pressure loss at height and to facilitate more reliable and efficient distribution of services. Engineers also have to carefully consider the space requirements of services to maximise the number of floors in a building and lettable area. In a mixed-use building, choosing the most appropriate heating, ventilation and air conditioning (HVAC) system for an office, bedroom or entrance lobby not only creates the best environment for occupants, but also leads to a more energy efficient and sustainable building.

As well as distributing services, tall buildings also have to distribute people efficiently. Advances in lift machinery and technology are improving the integration of these mini vertical cities, while also offering improved security and a means of controlling access to different areas. Using specialist simulation software, Arup analyses thousands of virtual journeys before deciding on the optimum lift configuration for a building, a process that has a profound effect on travel times and building revenue.

3a
© Wharf China Ltd

Case study 3a

Changsha International Finance Square
Changsha, China

CONTRIBUTORS: CALVIN FU, ERIC LEUNG + MING-YIU CHUNG

Client:	Wharf China Development Ltd
Design architect:	Wong & Tung International Ltd and Benoy Ltd
Executive architect:	Wong & Tung International Ltd
Arup's role:	Mechanical, electrical and public health engineer
Year of completion:	2017
Height:	452m (Tower 1) and 315m (Tower 2)
Number of storeys:	95 (Tower 1) and 65 (Tower 2)
Gross floor area:	Whole project including podium, 1,000,000m^2; Tower 1, 300,000m^2; Tower 2, 175,000m^2
Use:	Retail, office and hotel
Main contractor:	China Construction Second Engineering Bureau Ltd

Since the mid-nineteenth century, tall building design has evolved hand in hand with advances in lift design to allow architects and engineers to scale greater heights with smarter and more efficient vertical transportation (VT) systems. They are needed not just to move people around, but to also solve operational, maintenance and security issues within the integrated vertical cities.

Serving the vertical community

Changsha International Finance Square (IFS) in Changsha, the capital of Hunan province in southern China, is a good example of how VT plays an instrumental role in making large, mixed-use commercial developments successful. The development, with building GFA of 1 million m^2, is expected to host the tallest building in the province when completed and comprises offices, hotels, shops and leisure facilities across a podium and two towers.

The podium is made up of a five-storey basement with two mezzanine levels plus seven storeys above ground. Extruded from the podium are two towers – the 95-storey Tower 1 (T1) and 65-storey Tower 2 (T2). Offices occupy the majority of both towers, with a hotel occupying the top of T1 and the bottom

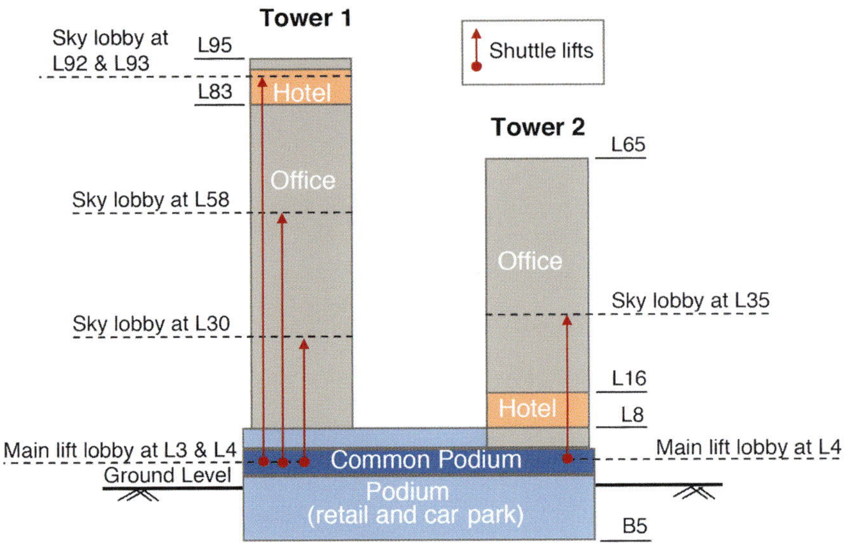

3a1
Schematic diagram of Changsha IFS
© Arup

of T2 (see Figure 3a1). The basement of the development will be connected to an underground railway station.

Mixed–use medley

There are many possible circulation routes for visitors to Changsha IFS due to its multiple functions and entry points. Its VT system rationalises these routes, offering clarity to visitors and security to building owners. For Arup, responsible for the design, this presented a number of challenges due to the height of the towers, the large population and the need to keep journey and waiting times short.

Had each tower only incorporated offices or a hotel, rather than both, the VT strategy would have been much more straightforward, but with both functions in one building, and due to the proportion of one occupancy type compared to the other, two separate lift systems were needed in each of Changsha IFS's towers – one for the hotel floors and another for the offices. T1, the taller of the two towers, will be discussed in more detail here.

Defining strategy

The solution at Changsha IFS, as in many tall buildings, is to incorporate sky lobbies. These are intermediate interchange floors higher up the building, and the only stopping point for shuttle lifts ascending from the main entrance lobby. These high-capacity, high-speed shuttle lifts allow people to travel quickly up the building before transferring to a "local" lift service which stops at every floor (until the next sky lobby), therefore reducing the total passenger transit time and the number of stops. The arrangement enables lifts to be stacked into zones, allowing multiple lifts to be operated within a single lift shaft.

3a2
The podium seen in the foreground has multiple entrances from street level, underground car park and metro
© Wharf China Ltd

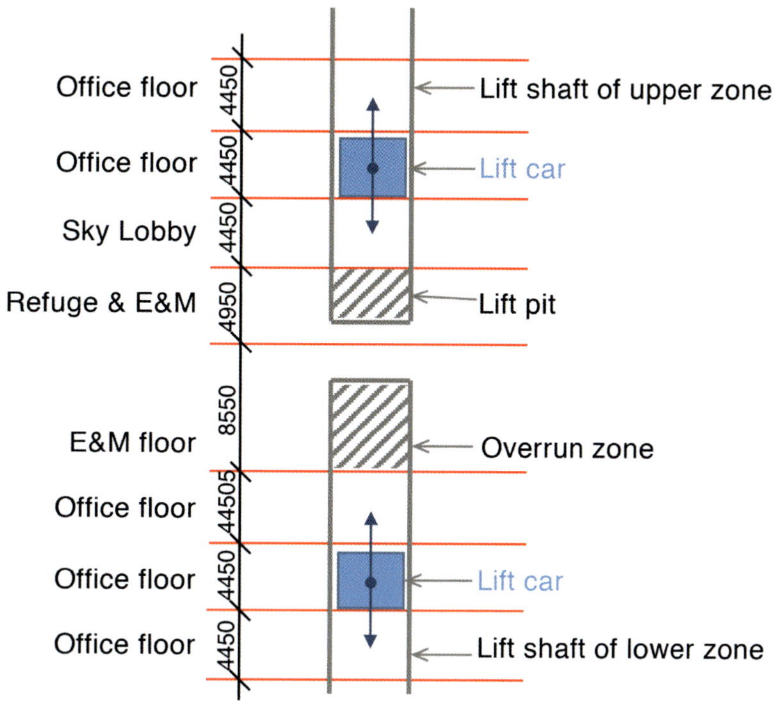

3a3
Location of the lift shafts, sky lobby and refuge floor under the stacking arrangement
© Arup

The sky lobbies need to be placed above the refuge and electrical and mechanical (E&M) plant room floors so that there is sufficient depth for the lift pit of the local lifts in that zone, and also to accommodate the overrun of the local lifts in the zone below (Figure 3a3). This limits the locations of the sky lobbies and the flexibility of the stacking arrangement. Arup's analysis revealed that two sky lobbies would be needed to serve T1's offices. As a result, the office portion was divided into three sections with sky lobbies at levels 30 and 58, above the refuge/

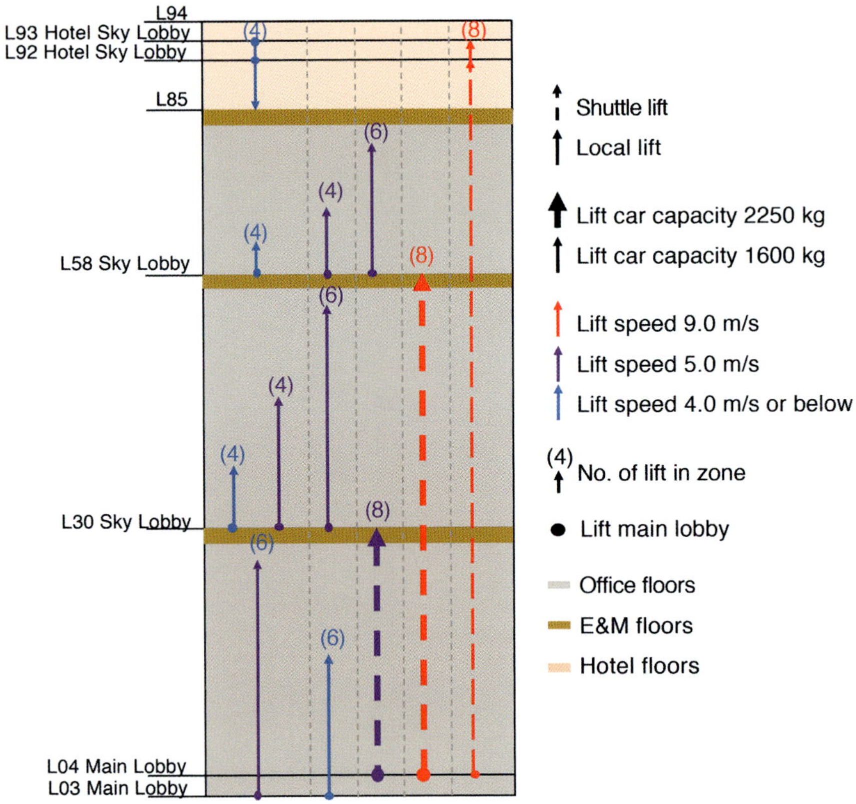

3a4
Conceptual diagram showing the lift stacking arrangement in T1
© Arup

E&M floors at levels 28–29 and levels 56–57, respectively. A separate shuttle lift zone with two sky lobbies at levels 92 and 93 will serve the hotel (Figure 3a4).

To achieve the shortest possible journey times, other strategies were also considered. For example, double-deck lifts, which take the form of one lift car on top of another were considered to be not justified in this project due to them requiring two floors of lift lobbies for every stop and additional space for escalators to allow transfer between their twinned lobbies. Single-deck lifts only require one floor for each sky lobby and provide a more efficient use of space, and are therefore the preferred solution in this project.

Optimising the journey

Lift design is based on the maximum number of people that will need to travel through the building at any one time, usually assumed to occur during the morning up-peak hour (8am–9am). For T1, predominantly an office building, the VT strategy was based on the client's requirement for an up-peak handling capacity of 11% – the percentage of the building's population that needed to be transported by lift during the most demanding five-minute period in the morning. This equated to 1,419 people.

See it, improve it

Lift traffic analysis software was used to compare different lift configurations that satisfied this handling capacity (11%) for different numbers of lifts, car sizes and travelling speeds. In addition, simulations were run to optimise the lift provision and verify that the lift sizes and speeds would be adequate for the maximum number of people travelling up the building in the morning. In each lift configuration, the program automatically generated 10 sets of random, but realistic, travelling patterns allowing deeper scrutiny of people flow and lift use.

The lift interval (the average time between successive car arrivals at the main lobby) of 35 seconds and 80% lift capacity were applied in the final VT design. Using the software, a single hour was scrutinised in five-minute intervals and it was confirmed that the lift interval and waiting time in all lift zones could meet the design requirements.

Further criteria such as comfort were also examined. A 3D computer model ensured that the lift lobby could support 70 people waiting for the lifts (the maximum queue length determined by simulation results) and would not be too crowded (Figure 3a7).

The lift traffic analysis and simulation software made investigating different configurations simpler, especially when structural or architectural changes impacted on the lift core. This was the case when the number of office floors increased by eight during design development. The number of lifts served within each zone and lift speeds were quickly altered to absorb these extra passenger loads without requiring additional lifts.

In all, 64 separate passenger lifts will be required for T1, stacked vertically in "zones" to suit the office and hotel arrangements and the shuttle and local lift

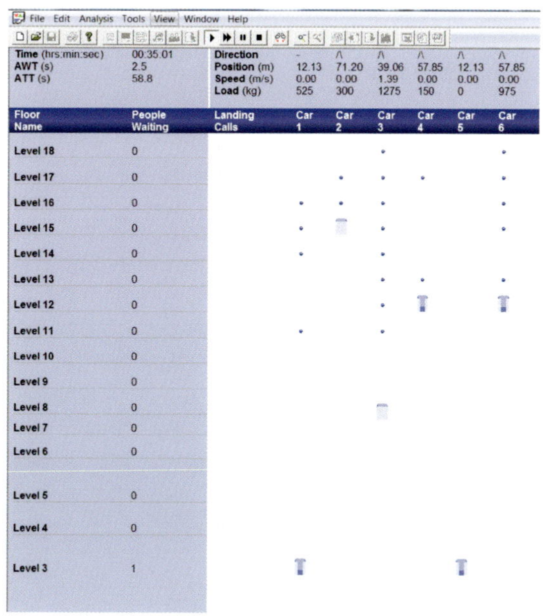

3a5
Lift simulation (screenshot) showing "movement" of lifts
© Arup

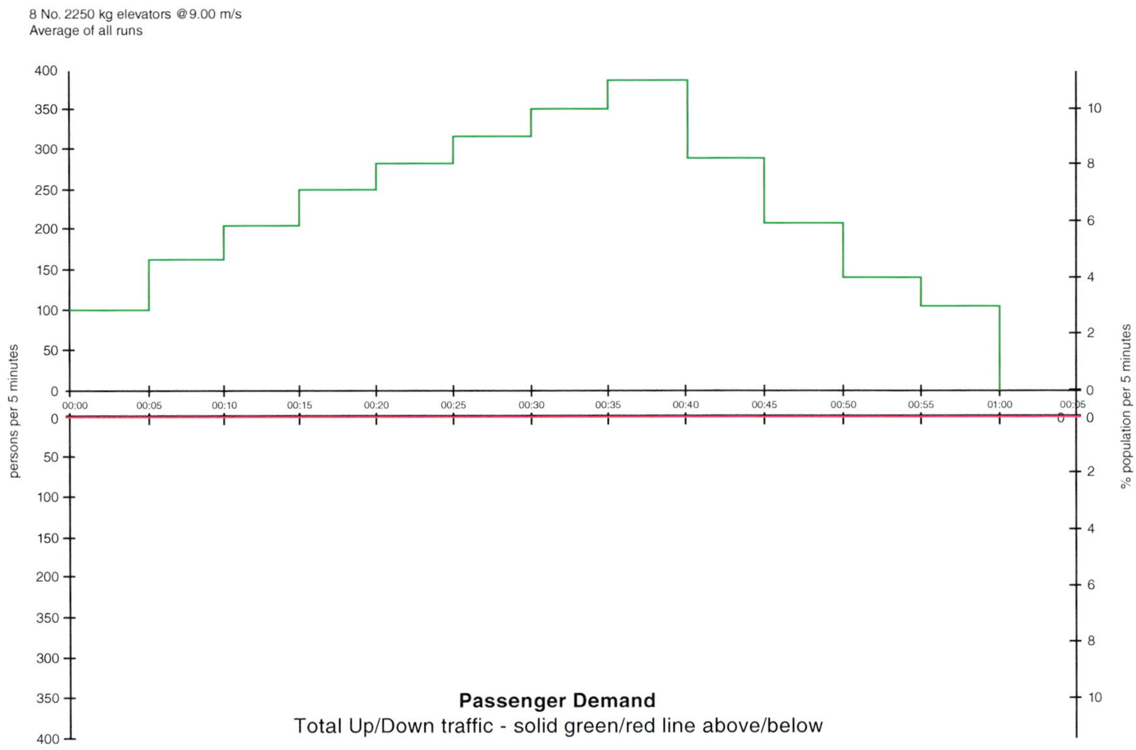

3a6
Lift simulation results (showing closer to 11% peak population served)

requirements. All lifts will be 1,600kg capacity cars except the shuttle lift cars for the offices, which are 2,250kg. Lift speeds will vary from 2.5m/s to 5m/s for the local lifts, and will be 5m/s or 9m/s for the shuttle lifts. Arup's VT strategy limits the average time to destination (the sum of passenger waiting time and transit time) to less than 100 seconds, ensuring the lift system can meet the 5-star service level according to the Chartered Institution of Building Services Engineers (CIBSE) Guide D. The stacking arrangement also reduced the building core size by 25% since 12 groups of lifts are derived from the plan area of just five lift shaft banks, freeing up more lettable floor area for offices.

Having two independent lift systems for the hotel and offices will also offer operational and maintenance advantages for both. Turnstile barriers could be installed in front of the privately owned office lift lobby to increase security, while a more personal approach using a reception desk could be used for the hotel. And since the two lift systems can be maintained separately, should any repairs need to be carried out they could be done during the day in the hotel and in the late evening in the offices when lift use would be low.

Visitors to Changsha IFS may never appreciate its lift system quite as much as its high-end hotel, shops and offices, but the fact that journeys will never be overcrowded and people will never have to wait too long for the next lift reinforces the development's aura of sophistication, one that exudes comfort, luxury and efficiency.

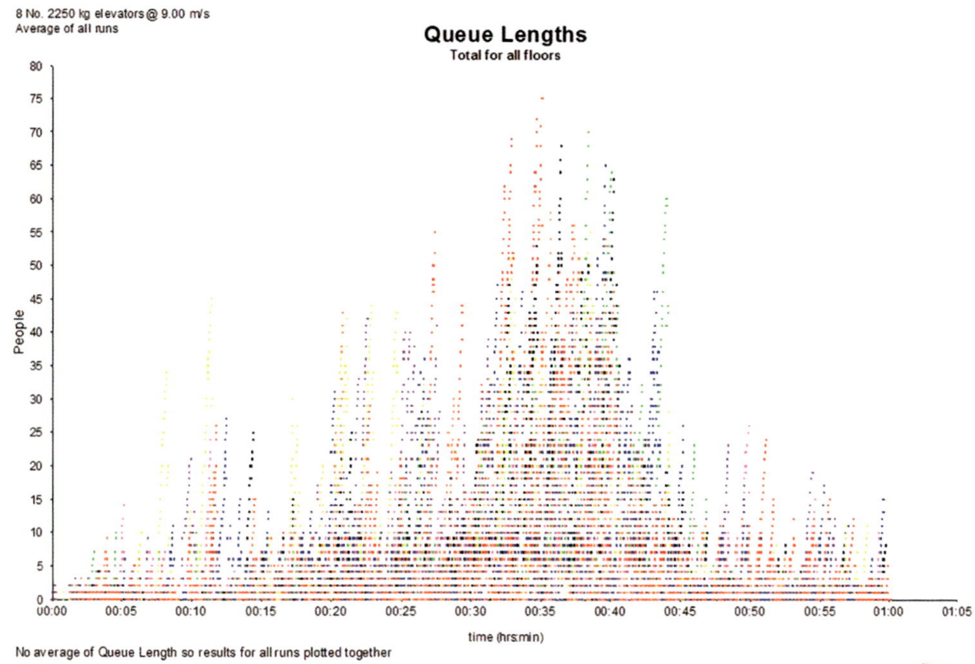

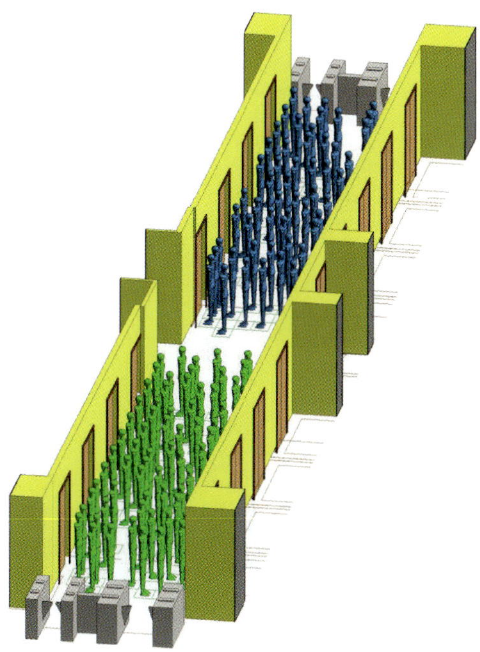

3a7
Simulation results showing queue length (top) and 3D computer model showing people waiting in the lobby (bottom)
© Arup

3b
© Zhou Ruogu Architecture
Photography

Case study 3b

Guangzhou International Finance Centre
Guangzhou, China

CONTRIBUTORS: YU-LUNG CHENG, TOM CHEUNG + BARRY LAU

Client:	Guangzhou City Construction and Development Group Co. Ltd
Design architect:	Wilkinson Eyre Architects
Architect of record:	Architectural Design and Research Institute of South China University of Technology
Arup's role:	Mechanical, electrical, public health, structural and fire engineer, façade, transport planning, vertical transportation and building sustainability consultant
Year of completion:	2010
Height:	438.6m
Number of storeys:	103
Gross floor area:	250,000m^2 (tower only)
Use:	Office and hotel
Main contractor:	China State Construction Engineering Corporation; Guangzhou Municipal Construction Group JV

The Guangzhou International Finance Centre (IFC) is a unique addition to Guangzhou's central business district. With a robust concrete-filled steel tube diagrid frame and reinforced concrete core, it requires no further structural armour – no belt trusses, outrigger or ancillary columns – to stretch 439m skywards. The simple purity of its form belies the intricacy of its design and, in particular, its hidden network of building services.

A luxury hotel with a 100m-tall atrium occupies the upper portion of the tower, while offices occupy lower levels. Its floorplate is, unconventionally, triangular in plan with aerodynamically designed curved corners and sides to help the building function more effectively in high winds. Its longitudinal profile experiences its greatest girth a third of the way up.

From the outset, the brief was for a low carbon and sustainable building services strategy that met high standards for occupancy comfort. There was also a priority to instil flexibility in the design so that it could adapt to meet future needs, especially as climate change places more pressure on heating, ventilation and air-conditioning (HVAC) systems.

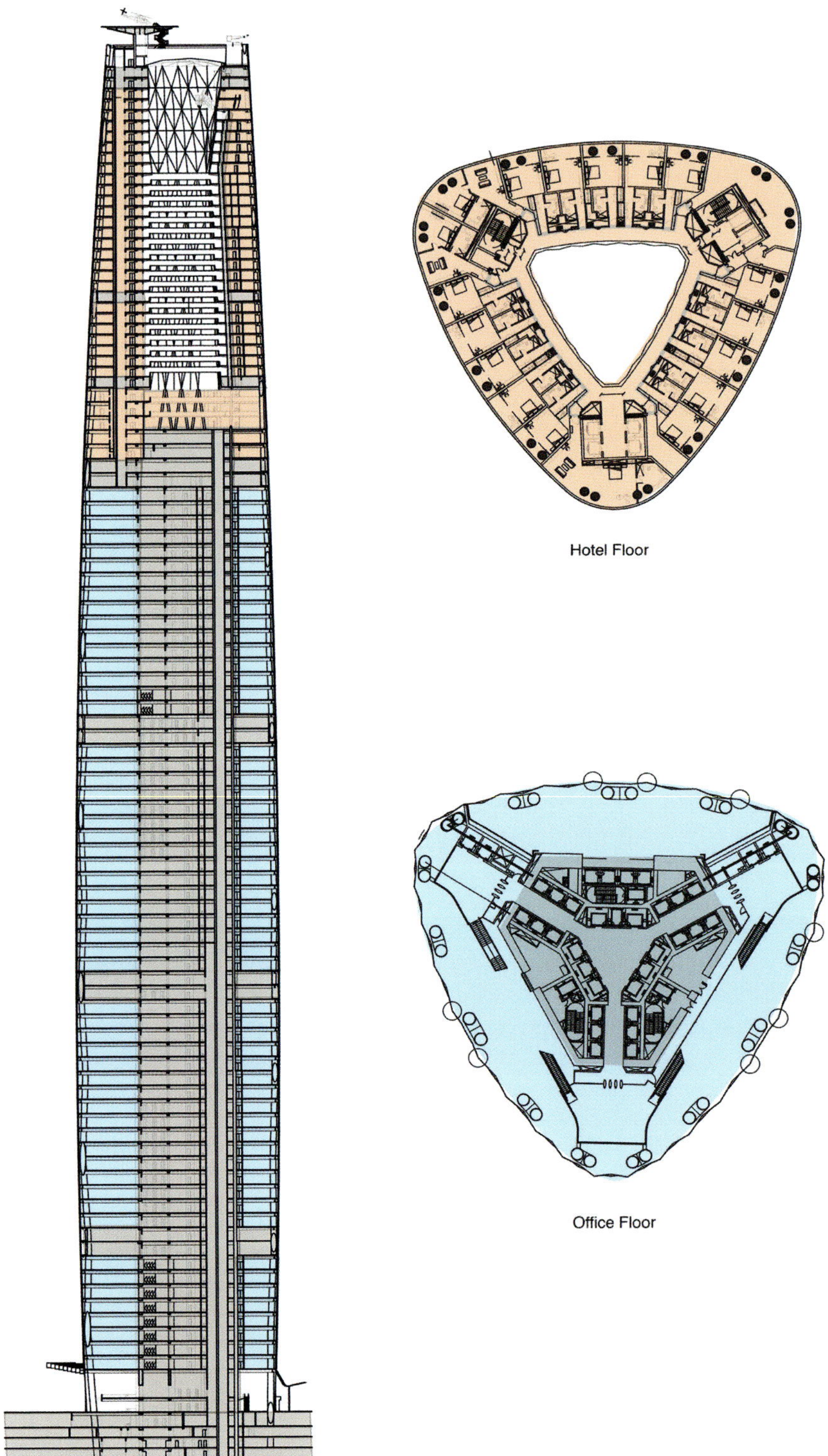

3b1
Elevation showing curved longitudinal profile and hotel and office floors, as well as plant level banding (left). Triangular floor plates with curved corners and sides show central core (bottom right) and atrium (top right) © Arup

Mixed-use medley

Guangzhou IFC's mixed-use nature inevitably meant that a different HVAC strategy was required to suit the different functions of the building. Offices need a significant amount of cooling by day, whereas hotel rooms, which are generally occupied from evening until morning, need much less cooling.

Guangzhou IFC's building services strategy consists of mechanical, electrical and plumbing (MEP) equipment housed on five designated plant levels spaced across the building's height. To save space, these plant levels are integrated with refuge floors that are used for fire evacuation. Each plant level serves the eight floors above and below it. Equipment is housed on different levels because co-locating at ground level, say, would pose too many constraints on valuable floor space.

The nature of tall buildings also means that services need to travel long distances against gravity to reach their destinations. By distributing MEP services throughout the tower, the distance travelled is reduced via few stages and the need of using high head/static pumps and fans to reach the farthest end of the building is avoided.

The different functional zones of the building are served by specific plant levels and HVAC systems, which allow simpler division of services to suit different tenant needs.

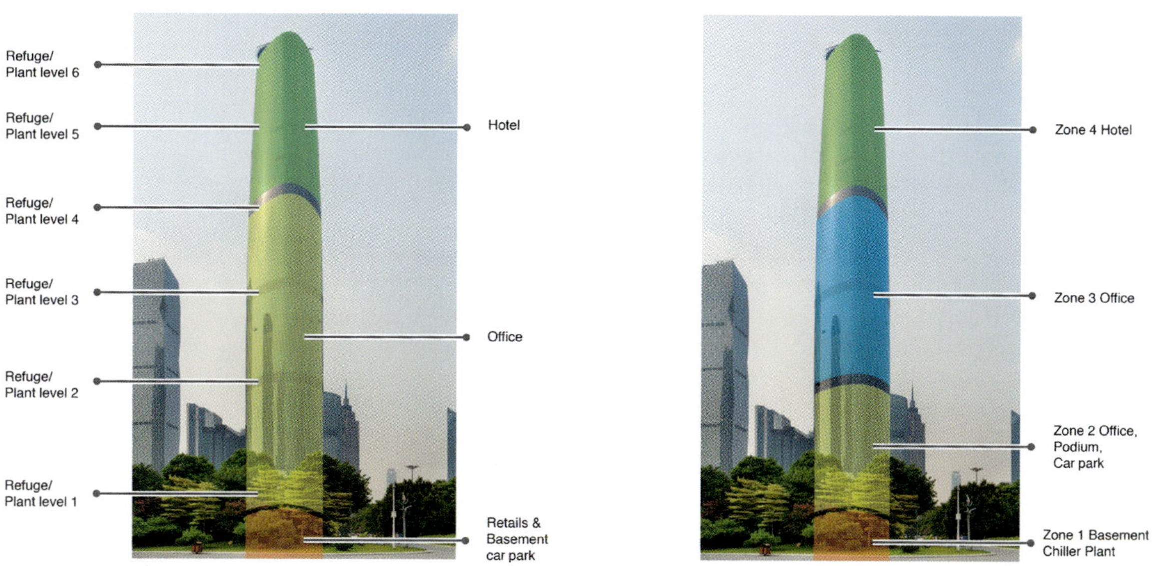

3b2

The building's functional zoning and refuge and plant levels (left) and MEP zoning strategy (right)
© Arup

3b3
The refuge floors and plant floors can be easily identified by banding on the façade (left), especially at night (right)
© Zhou Ruogu Architecture Photography

The core and ceiling conundrum

As with all tall buildings, there is a need to design the core to be as compact as possible to free up more lettable floor space to achieve higher floor efficiency. The majority of a core is taken up by lifts, staircases and washrooms, but they also accommodate plant rooms and risers where services travel up the building. Exacerbating the task of cramming all these items into Guangzhou IFC's core is its triangular shape, which adds a geometrical constraint to its design.

Services are distributed from the core to each floor via ceiling voids. But a balance has to be struck between how much space these services occupy in the core and how much they occupy in the ceiling. Equipment for HVAC systems that take up less space in the core typically occupies a greater depth in the ceiling void for distribution. A deeper ceiling services zone then leads to a decrease in clear headroom, which can be at odds with the architectural intent of the space. Achieving this balance involves extensive coordination between the building services and structural engineers.

Cool and in control

To maximise the clear headroom on each office floor, Arup explored different HVAC options, paying particular attention to duct and pipe sizes and efficiency. It

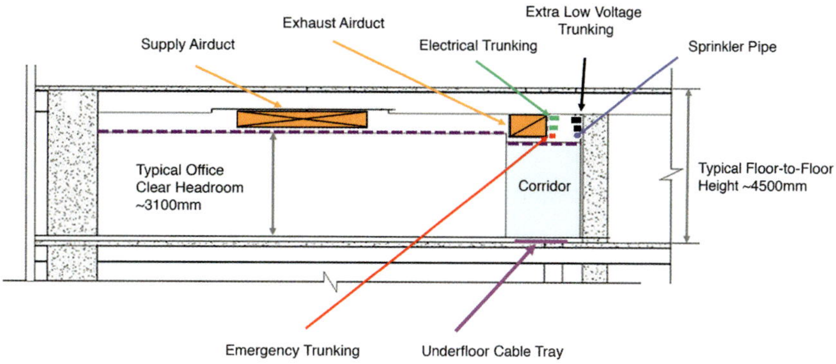

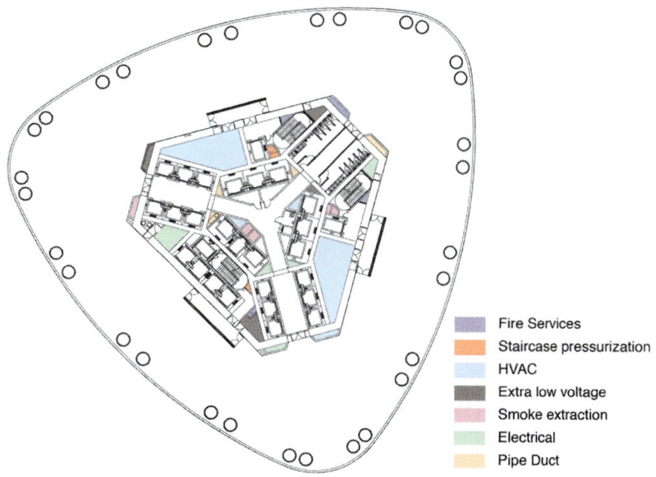

3b4
A typical floor showing services housed in the ceiling void (top) and core (bottom)
© Arup

compared the suitability of fan coil unit (FCU), variable air volume (VAV) and under floor air-conditioning (UFAC) systems.

Although an FCU system would achieve a slimmer ceiling depth compared with a VAV system, the latter's greater capacity for higher cooling loads and the avoidance of running chilled water pipes inside the ceiling void meant that it was more suitable for Guangzhou IFC's office floors. VAV systems supply air at a constant temperature and vary the volume of air supplied at part load conditions. So, if a thermostat identifies that the temperature is too high within a zone, the system is commanded to supply more air to that area to cool it down. Similarly, the volume of air supplied is reduced if the room is too cool. The mechanism by which VAV systems operate has a high degree of control and can lead to more direct and efficient heating and cooling than FCUs. A further disadvantage of FCU systems is that their maintenance requires accessing tenant space, which can be disruptive.

A decentralised HVAC distribution system was adopted, with each office floor equipped with two air handling units (AHU) so that cooling and ventilation could be controlled locally on every office floor to suit the needs of each tenant.

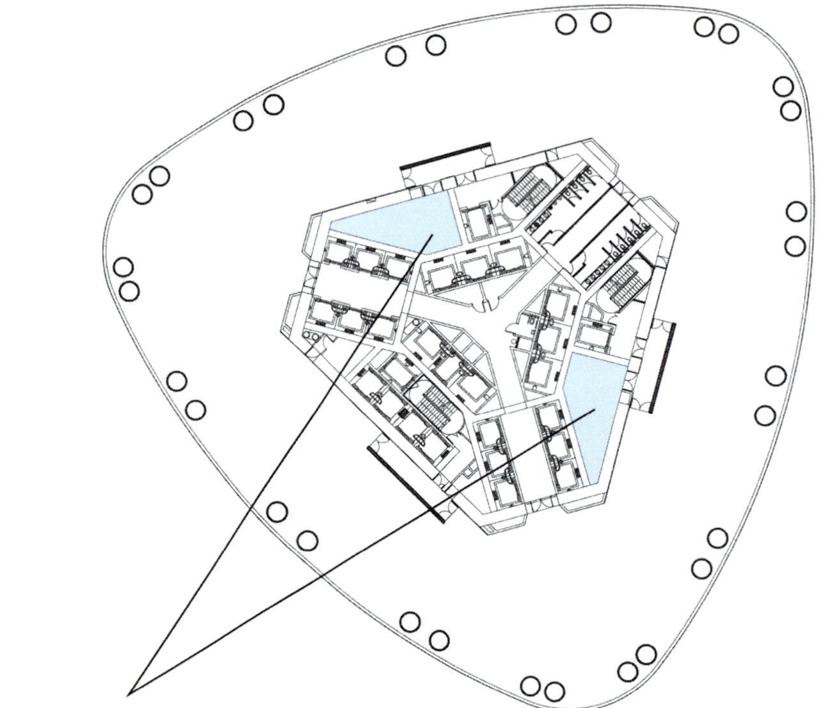

3b5
Two plant rooms within the core of each office floor of Guangzhou IFC
© Arup

The system includes primary air handling units (PAU) on plant levels supplying pretreated fresh air to two AHU rooms on each office floor. Ducts from the AHU rooms then pass through the ceiling void and deliver cool air into the offices on the same floor. The PAUs provide a heat recovery wheel to cool and dehumidify incoming fresh air before delivering it to the AHUs on the office floors.

The PAU design can achieve 50% maximum free air cooling which reduces the building's reliance on artificial cooling and reduces duct sizes. Although plant room space is taken up in the core using this system, it also offers flexible and easy access for maintenance and replacement in the future.

Having two AHU rooms and, hence, two VAV systems per floor allows one to serve half of the office spaces to minimise duct run and pressure drop. An interconnecting duct between the main duct of these two AHUs enhances the system's reliability. If one AHU malfunctions, the other can still supply a proportion of cool air to the tenant space.

Personalised service

On the hotel floors, an FCU system with pretreated fresh air was adopted. Here, the PAU is located "centrally" on plant levels and chilled water is piped to each guest room. A thermostat in each guest room regulates how much chilled water is required to maintain the desired temperature.

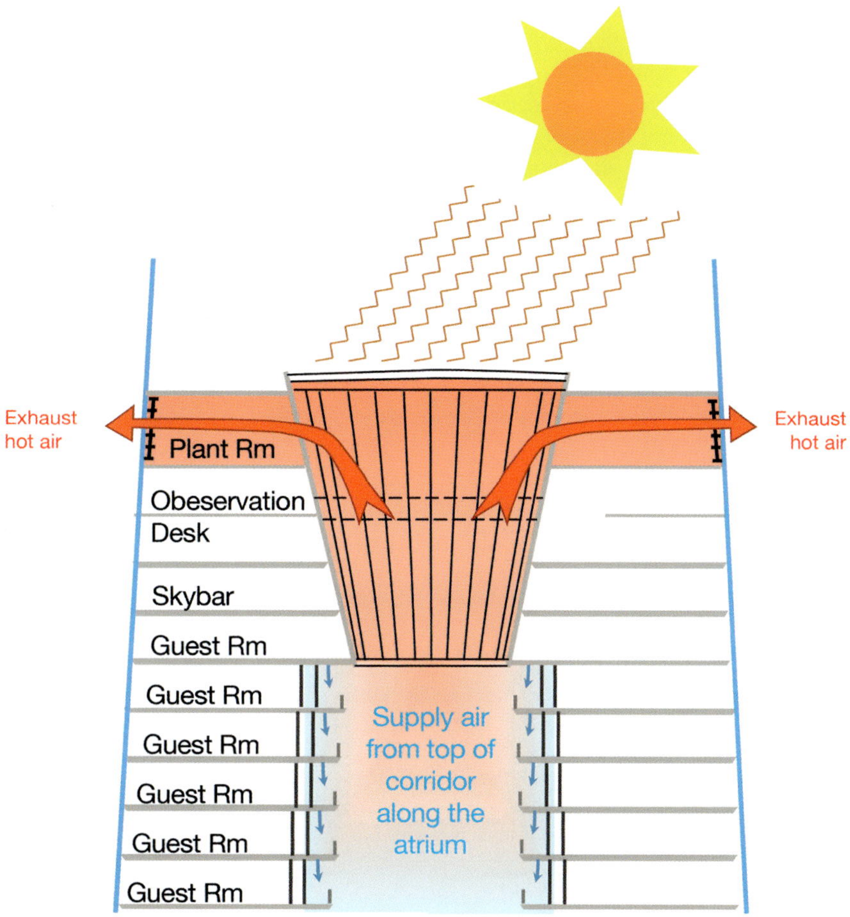

3b6
Air supplied at guest room corridor pushing the hot air in the atrium to exhaust at the high-level plant room
© Arup

Atrium highlights

Displacement ventilation, or under-floor air-conditioning is used in the hotel lobby and office lobby to reduce energy consumption and running costs. The system cools air up to 2m in height (just above the height of people, namely the occupied zone) and avoids wasting energy on heating or ventilating the entire space.

Airflow in the atrium was analysed in more detail to understand whether air would rise (mainly due to the solar heat gain at the top) and collect in any of the balconies overlooking it. Computer models identified that additional fresh air was needed in balconies to "push" warm air out.

Gone with the wind

The zone around the diagrid structure is used as a plenum to allow airflow around the perimeter of the building. One of the challenges of tall buildings is that exhaust air can clash with high winds blowing onto the face of the building more than 200m above the ground, making it difficult for this air to be expelled. At Guangzhou IFC, exhaust air enters this plenum and is drawn through on a current of air generated from the prevailing wind "pushing" on one side of the building and "pulling" on the opposite side. To ensure exhaust air does not mix with intake air on any one side, they are placed 10m apart.

Guangzhou IFC is a building whose architecture purposefully does not reference the heritage of the land it sits on, focusing instead on having a form and function that will stand the test of time. This is also true of its building services, which also transcend current trends, choosing instead to place the design emphasis on efficiency, sustainability and flexibility to meet ever-changing needs.

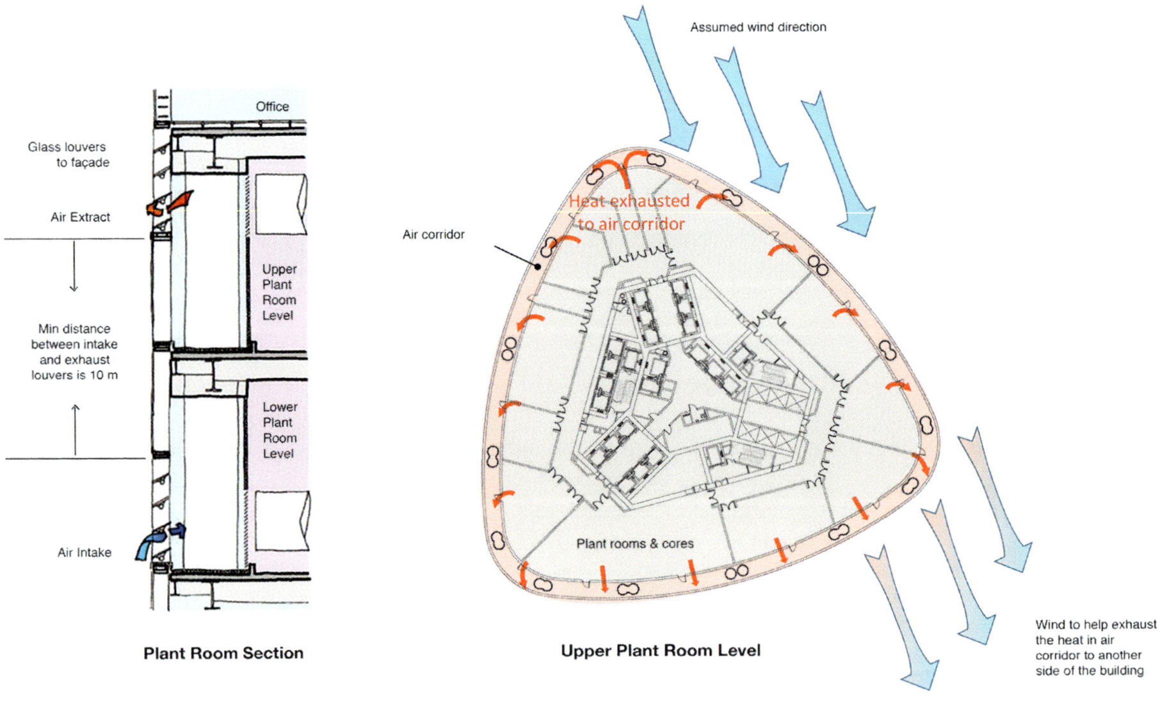

3b7

Air intake and extract louvres on the same side of the building are at least 10m apart in elevation (left). They create a current of air which draws exhaust air away from where wind is blowing onto the face of the building (right)

© Arup

Chapter 4

Safe and comfortable

Events such as the World Trade Center collapse in New York in 2001 highlighted safety concerns unique to tall buildings. To better protect occupants, engineers, so used to being concerned about how these buildings stand up, now also had to understand much more about how and when they might fall down.

Using specialist computer software, historic data and practical engineering rationales, Arup works at the forefront of structural fire engineering to understand how tall buildings behave in extreme situations, producing designs that exceed the requirements of current guidance.

This degree of scrutiny goes hand in hand with examining the practical aspects of moving tens of thousands of people safely to street level in an emergency. By looking at a range of solutions and the mobility needs of a more diverse population, Arup is designing buildings where lifts can complement the use of stairs to offer a fast and reliable method of evacuation.

Where typhoons and earthquakes occur habitually in many regions of East Asia, the motion they induce in tall buildings must be controlled so that occupants feel safe and comfortable and business can continue. Designing a building to be more resilient can be the driver for some wonderful innovations, including Arup's audacious plan to release floors from the main frame of one building to counter the sway caused by seismic activity.

4a

Case study 4a

China Zun (Z15)
Beijing, China

CONTRIBUTORS: YOUNG WONG + VALA YU

Client:	CITIC HEYE Investment Company
Concept architect:	Terry Farrell and Partners
Design architect:	Kohn Pedersen Fox Associates
Associate architect:	Beijing Institute of Architectural Design CITIC General Institute of Architectural Design & Research Company
Arup role:	Fire, structural and geotechnical engineer
Year of completion:	2018
Height:	>500m
Number of storeys:	108
Gross floor area:	437,000m^2
Use:	Office
Main contractor:	China State Construction Engineering Corporation

Located in the heart of Beijing's central business district is the China Zun Tower, a 500m plus tall office building, which will accommodate more than 20,000 people when it is completed in 2018. Inspired by the shape of a traditional Zun water vessel with its characteristic slim waist, and enlarged opening and base, the building is intended to be a symbol of communal goodwill.

Fundamental to the design of the tower's perimeter steel frame structure is the desire to facilitate the safe evacuation of its occupants in the event of a fire. The Zun is one of the first high-rise buildings in China to use structural fire engineering analysis for the design of its main structural system. The structure has also been designed to resist earthquakes, high winds and gravity loads, as would be expected for a building of this height and in this location.

Tall building design post-9/11

How a building will respond in an emergency, such as a fire, and the means of evacuation became of greater concern for tall building clients and their designers following the collapse of the World Trade Center towers 1, 2 and 7 in New York on 11 September 2001 ("9/11").

Recommendations for tall building design made by the US National Institute of Standards and Technology (NIST) following 9/11 go beyond that prescribed by national codes. They highlight that interaction between different structural members can be beneficial or detrimental to a building during a fire.

Hence, the fire response of "super-tall" buildings such as the Zun, with its "mega" members and connections, have to be justified on a case-by-case basis to ensure they offer adequate protection to occupants and surrounding areas.

Limitation of existing codes

Chinese fire codes state that a building needs to remain standing for at least three hours to allow complete evacuation. To achieve this level of fire resistance, it suggests that for column members when tested to standard fire curve, main gravity load transfer members should have a 50mm thick applied protective coating. However, the guidance is limited, as it considers each member in isolation and disregards the interaction between adjacent members or whole building behaviour.

This interaction can be beneficial to a building because adjacent members can help support a fire-damaged member by offering different load distribution paths. But it can also be detrimental, causing early buckling of members due to adjacent members restraining thermal expansion. The exact impact of this interaction can be understood in greater depth by carrying out a finite element (FE) study.

Arup also considered the 50mm thickness of fire protection coating excessive for the Zun, where some steel members have increased inherent fire resistance as they are concrete-filled. The consultant was also concerned that 50mm of coating on all members would increase the weight of the building and increase costs. FE analysis into how the structure would respond during a fire took into account the most likely fire scenarios based on real fire events. These fires lasted longer and involved more floors than specified in design codes.

The Zun's structure

The 108-storey Zun tower sits on top of a seven-storey basement and consists of a central reinforced concrete core and perimeter steel braced frame. This perimeter steelwork is made up of gravity columns (which support gravity loads only), belt trusses (which visually divide the height of the building into seven zones and form part of the overall lateral stability system), mega columns (which run the full height of the building) and mega cross-bracing (Figure 4a1).

The building is square in plan and narrows from 78m at the bottom to a 54m wide waist, enlarging again to measure 69m across at the top. The narrowest portion of the building is 385m from the base.

4a1
The Zun structural system
© Arup

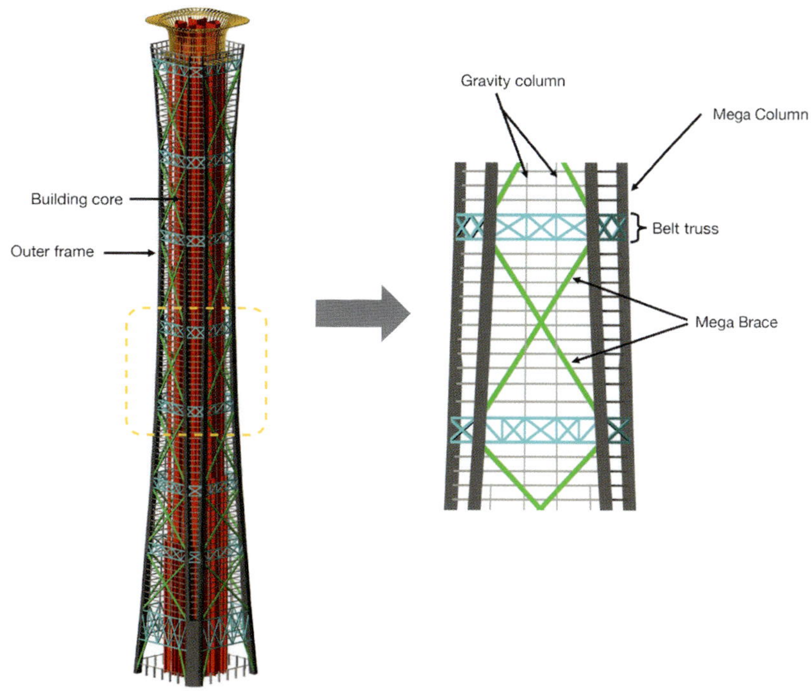

4a2
First layer of mega brace and belt truss
© Zhou Ruogu Architecture Photography

Simulation of building fire resistance

Computer simulations were also run to estimate the time needed to evacuate the Zun. A well-organised simultaneous evacuation of 20,000 occupants would take two hours, predicted Arup. However, in reality, many occupants would take much longer to make the descent to street level, due to possible anxiety, injury or poor health.

This indicated to Arup that the structure had to remain sufficiently intact and resist progressive collapse during a fire. Progressive collapse is where failure of a single structural member or connection causes, say, a whole portion of the building to fail and that the excessive unsupported weight of this causes subsequent failures, leading eventually to the collapse of the whole building.

With all this in mind, detailed structural fire analysis of the Zun using FE modelling was undertaken to determine exactly how it would respond during a fire (Figure 4a3).

Analysis revealed that a steel reinforced concrete mega column with 6mm of fire protection (Figure 4a4) would not be detrimentally affected by high temperatures even after six hours of burning (based on standard fire curve) because the majority of its concrete centre is unaffected by the heat.

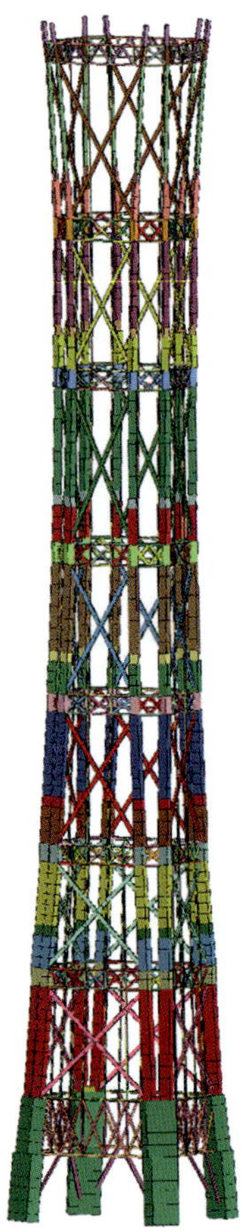

4a3
FE model of the perimeter steel frame
© Arup

4a4
Heat distribution (right) across a steel reinforced concrete mega column (left), modelled as a quarter of a column cross-section with 6mm of fire protection, after 6h in a fire
© Arup

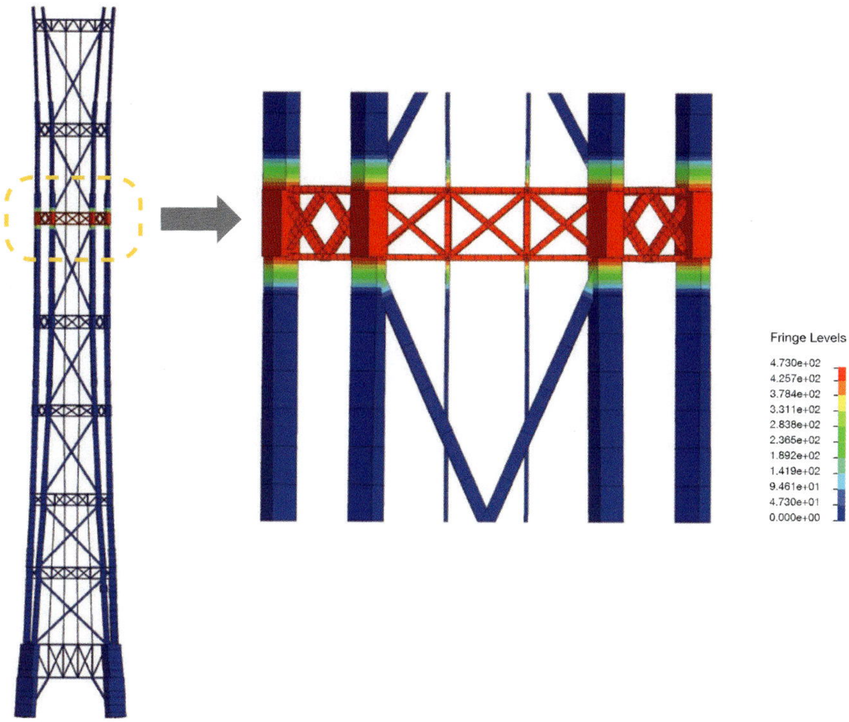

4a5
Fire analysis of the belt truss assumed that a worst-case fire took place on the two floors directly above zone 5. This particular belt truss (red) was modelled as it was made up of smaller structural sections compared to others in the building
© Arup

Consideration of thermal restraint effects

Arup then went on to investigate the effect of a fire on critical parts of the structure. This exceeded Chinese design code requirements, but took on board NIST's recommendations and also concern from Chinese authorities in light of a recent building fire in Beijing.

The analysis revealed that the belt truss was the most vulnerable to thermal expansion in the structure, because its diagonal members would be subjected to very high axial restraint (Figure 4a5). Arup assumed that an entire belt truss was on fire in the FE model. The thickness of fire protection on the belt truss, mega columns, mega bracing and gravity columns was varied in the FE model and their fire response evaluated. Their effect on the rest of the building was also considered.

The FE model verified that during a credible worst-case fire, thermal expansion of belt truss members would be restrained by the rest of the building and that extensive buckling of some structurally less important members would occur, but that key load-carrying members would be able to remain intact for longer than had previously been assumed.

These results also demonstrated that more than three hours of fire resistance under standard fire exposure could be achieved using just 6mm of protective coating on the mega column, 10mm on the belt truss/mega bracing and 20mm on the gravity columns (Table 4a1).

Table 4a1 Fire protection required to achieve more than three hours of fire resistance

Members	Calculated Protection Thickness	Required Protection Thickness	Calculated Fire Resistance	Required Fire Resistance
Mega column	6mm	50mm	>4.9h	3.0h
Belt truss/mega bracing	10mm	50mm	>6.0h	3.0h
Gravity column	20mm	50mm	>6.0h	3.1h

Analysis of critical connections

Arup assessed all major connections to determine the tendency for progressive collapse to occur as a result of a fire. The bulky connection between the mega column and bracing (or belt truss) was designed more critically for earthquakes compared to fire conditions.

The bolted connection between gravity columns and the lower chord of the belt truss was, however, examined in detail, with the assumption that its bolts would fail in a worst-case fire (Figure 4a6). The analysis demonstrated that, despite two gravity columns becoming disconnected from the belt truss and deformation due to expansion in the steel members, the whole structure remained stable even after six hours of burning.

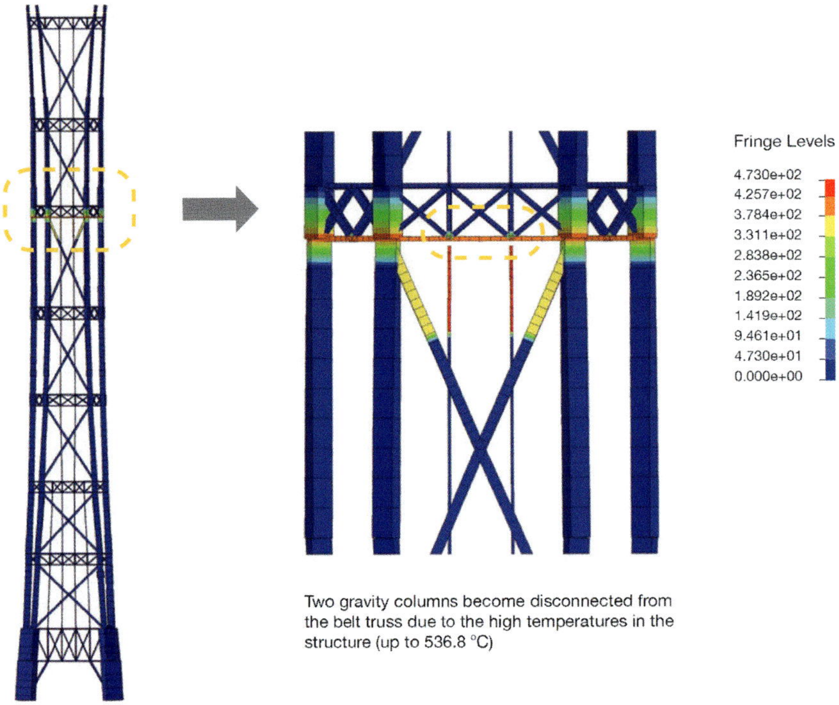

Two gravity columns become disconnected from the belt truss due to the high temperatures in the structure (up to 536.8 °C)

4a6
FE analysis of connections during a fire revealed that the bolted connection between gravity columns and the lower chord of the belt truss was most critical. The FE model shows the temperature distribution after six hours of burning and verifies that the building would still be standing © Arup

The benefits

Arup's structural fire analysis of the Zun validated that the thickness of fire protection coating could be reduced significantly, offering significant cost savings.

Arup demonstrated that the entire structure would remain stable for six hours during a worst-case scenario fire and offered a deeper understanding of how the structure would behave: while some members might fail, their loss would not cause the whole building to collapse and would allow enough time for evacuation and for firefighters to enter and leave the building safely.

The Zun's six-hour fire rating is far in excess of the three hours stipulated in Chinese design codes.

4b
© Daichi Ano

Case study 4b

Nicolas G. Hayek Center
Tokyo, Japan

CONTRIBUTOR: RYOTA KIDOKORO

Owner:	Swatch Group
Client:	Shigeru Ban Architects
Architect:	Shigeru Ban Architects
Arup's role:	Structural engineer
Year of completion:	2007
Height:	56m
Number of storeys:	14
Gross floor area:	5,679m^2
Use:	Retail and office
Main contractor:	Kajima Corporation, Suruga Corporation

The Japanese flagship store for Swiss watch manufacturer, Swatch, is as intricately and assuredly designed as the timepieces it displays. This 14-storey building located in Ginza, Tokyo's high-end shopping quarter, includes three mechanical features that set it apart from others.

Firstly, the seven Swatch brand floors in the building are reached via showroom elevators – entire rooms that move vertically to deliver customers to dedicated shopping levels. Then there are the three- or four-storey retractable glazing walls that allow fresh air to flow and create a public thoroughfare at ground floor level. Lastly – and making this building unique – is the way it mobilises the mass of four upper floors to counteract the building's sway during an earthquake.

Called the Nicolas G. Hayek Center after the Swatch Group's founder, the client wanted the building to be designed to the most onerous seismic design standards, known in Japan as "special grade". This meant that even after a one-in-a-1,000-year earthquake, it would not collapse. The integrated solution that meets this standard can be found on floors 9, 10, 12 and 13. These floor plates are dislocated from the main structural frame and instead rest on bearings that allow their mass to be released to protect the building from earthquake damage.

Developing this "self-mass damper" (SMD) system was the result of an extensive study into how earthquake resistance, which is traditionally provided by a beefed-up structure or space-gobbling dampers, could be incorporated into

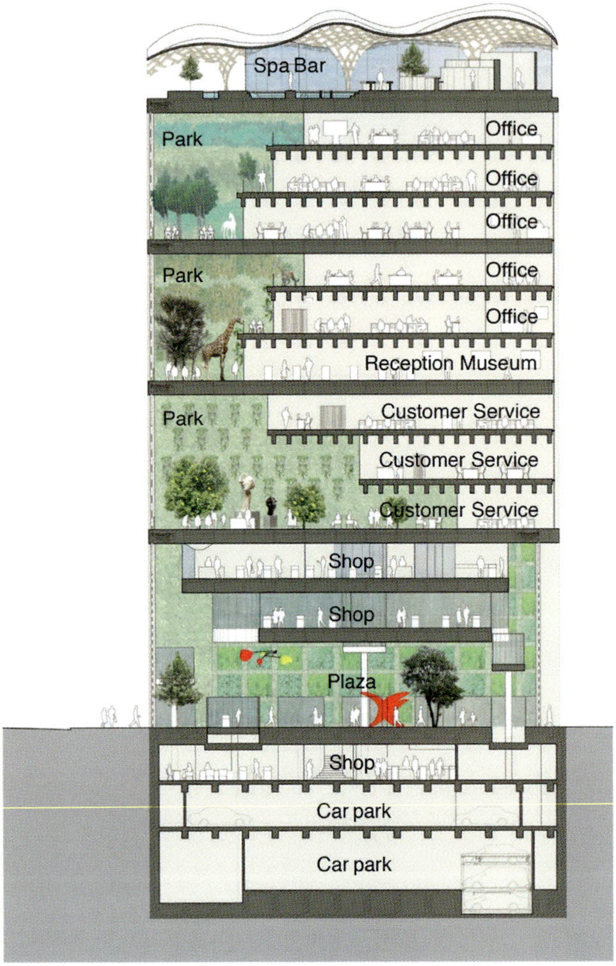

4b1
Cross-section through building showing shop floors and sky gardens
© Adapted from Shigeru Ban Architects

a mid-rise, narrow building with multiple atria, standing shoulder to shoulder with adjacent buildings.

Main structure

The Nicolas G. Hayek Center is steel-framed with a 12.6m wide elevation and 31.2m deep floorplates. Its rigid steel structure, which is massed into three-storey chunks (four-storey at the ground floor) is clearly identifiable from the building's front elevation, and is repeated at 2.4m intervals along its depth. A three-storey atrium extends from front to back at ground floor level to create a sense of space in the narrow building. Elevator showrooms glide up and down this space and, when the front and rear glazed elevations are opened up, a new public street is created. To compensate for the scarcity of continuous structure across the main atrium, extra bracing was introduced around the stair cores.

4b2
Ground floor atrium showing showroom elevators and hanging floors
© Daichi Ano

Three retail floors hang from the underside of the fourth floor and overlook the atrium, while offices occupy the levels above, each with a triple-height sky garden facing the street. In all, ten lifts, including a car lift and two stairwells, pass through the ground floor to service the building, which has resulted in the ground floor slab requiring significant strengthening using 6–12mm-thick steel plate.

Three-storey moment frames at levels 8, 11 and 14 stiffen up the building, while the intermediate floors are home to the building's SMD system.

Exploring damping mechanisms

The starting point for the Nicolas G. Hayek building's seismic design was to look at the suitability of damping mechanisms currently on the market. Base isolation was the most obvious solution – where the building is placed on rubber bearings and allowed to slide up to 1m away from its rest position during an earthquake. But such a system would eat into valuable retail space, so was rejected. Installing hysteretic or viscous dampers in the three core frames was also explored, but was found to fall short of the seismic response needed.

Mass damping mechanisms require some degree of mass to be mobilised to counter movement – base isolation uses the weight of the entire building – but adding extra weight purely for damping was considered counterintuitive, since the structure would then need to be beefed up, jeopardising the architectural intent for a light, airy building.

Instead, engineers looked at ways in which only part of the building provided the mass needed to dampen its oscillation during an earthquake. The notion of hanging two floors to allow them to "swing", much like a pendulum, was momentarily entertained, but would ultimately cause too much disruption to the way these floors could be used and so was not viable.

The final solution brought together the concepts of base isolation and swinging pendulum and involved supporting a number of floors on bearings mounted on stubs or corbels instead of the main structure. The mass of the floors could then be mobilised – to slide more than swing – without impacting on the main structural frame. By accommodating a modest gap around each floor plate to allow for this movement, there would also be minimal impact on available retail space. The task now was to find a bearing suitable for the job.

Developing the bearing solution

Working closely with bearing manufacturers, Arup adapted conventional base isolator bearings to suit the particular requirements for the Nicolas G. Hayek Center. Such bearings are typically designed to resist the weight of a building and contain steel plates to prevent the rubber damping material being crushed. Since the vertical loads on the bearings at the Nicolas G. Hayek Center would only be from a single floor, the plates could be removed and the rubber itself was designed to suit the required lateral damping criteria (Figure 4b4). Additional slider bearings would be added to create a system to resist gravity loads.

The rubber material's composition was carefully formulated to be stiff enough to provide adequate resistance and damping for a one-in-a-1,000-year earthquake, but also sufficiently pliable enough for more extreme earthquakes to allow stresses to be released. Five pairs of damper bearings were arranged on each floor along with four pairs of low friction slider bearings. The latter supported the weight of each floor and allowed

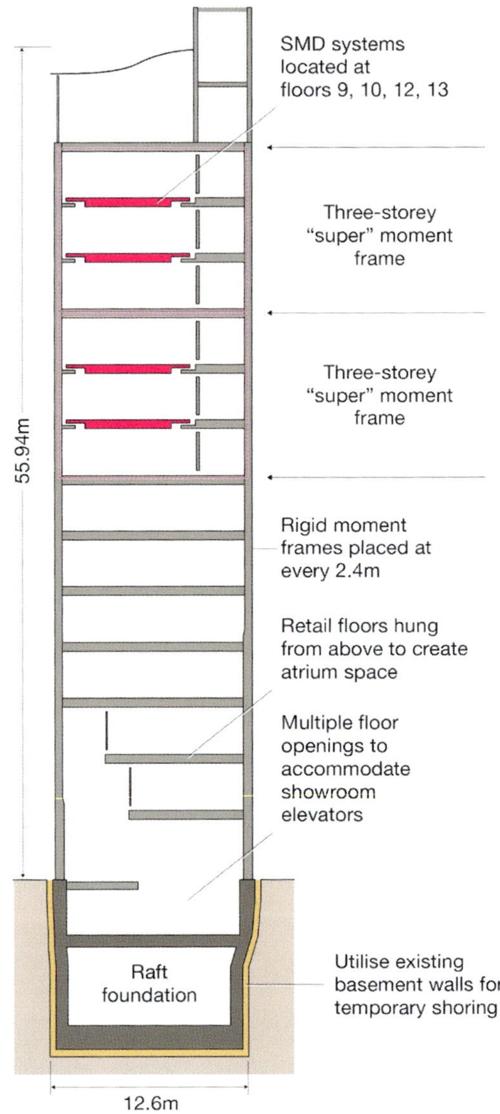

4b3
Building cross-section showing moment frames, SMD floors and atrium
© Nigel Whale/Arup

4b4
Rubber bearing testing (no steel plate within depth of bearing)
© Arup

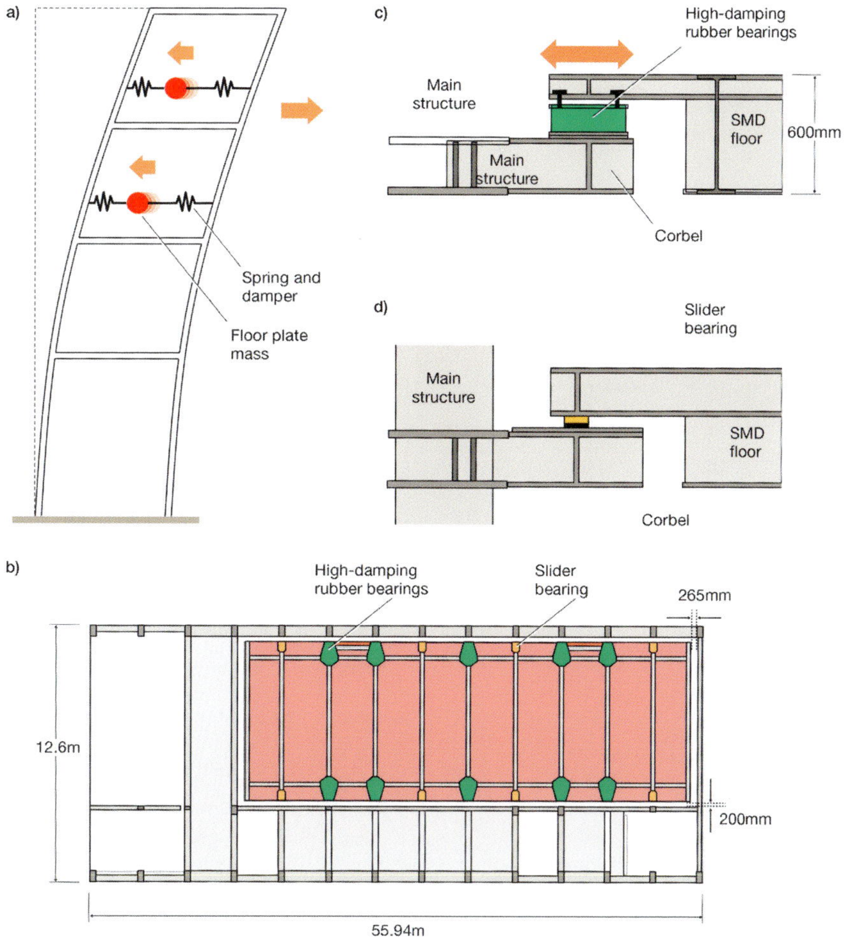

4b5
The "self-mass damper" system: (a) concept, (b) floor plan, (c) section showing rubber bearing and (d) slider bearing
© Nigel Whale/Arup

lateral movement. All bearings were less than 150mm thick and, along with the corbel depth, could be accommodated within the 600mm floor beam thickness (Figure 4b5).

With each floor weighing 100t, four floors contributed 400t or approximately 10% of the superstructure's weight – equivalent to the mass required to provide adequate damping. The bearings were tuned to provide maximum damping for the overall structure while limiting lateral movement of the floor plate itself to 200mm in the short direction and 265mm in the long. Mechanical and electrical fittings had to be detailed carefully across this variable gap to ensure they would be unaffected by the floor movement. Computer analysis and large scale testing confirmed that the SMD would reduce building movement by up to 35% in a major earthquake.

This unconventional, yet simple concept for controlling seismic motion ensures that floor space in the building is maximised and its light and open areas can be realised. With a living "green" wall rising up through the structure and retractable glazing allowing fresh air to circulate, the Nicolas G. Hayek Center is a welcoming building, which brings the outside in. Underpinning this character is the confidence to take on the forces of nature too; where the structure uses its own muscle and flexibility to temper the force of earthquakes.

4c

Case study 4c

Shanghai World Financial Centre
Shanghai, China

CONTRIBUTORS: MINGCHUN LUO, YOUNG WONG + KELVIN WONG

Client:	Mori Building Co. Ltd
Design architect:	Kohn Pedersen Fox Associates; Mori Building; Irie Miyake Architects and Engineers
Associate architect:	ECADI; Shanghai Modern Architectural Design Company
Arup's role:	Fire engineer
Year of completion:	2008
Height:	492m
Number of storeys:	101
Gross floor area:	381,600m^2
Use:	Hotel and office, observatory, retail, restaurant and parking
Main contractor:	China State Construction Engineering Corporation; Shanghai Construction Group

As tall buildings get taller and the trend for mixed-use developments continues, engineers and architects are having to produce designs which are more inclusive of the broader needs of building users. These vertical cities, containing a mix of private offices and apartments, as well as hotels, restaurants, shopping malls and leisure facilities, have always been designed for different floor loadings and dynamic responses. But as their facilities become more accessible to the general public, designers are also having to take into account the mixed abilities of occupants – the mobile and less mobile; the young and the old.

Since the collapse of the World Trade Center buildings in New York in 2001 ("9/11"), emergency evacuation of tall buildings has been analysed in more detail to develop better solutions than those provided in current design codes, which currently fall short of addressing the unique issues of mixed-use tall buildings.

Current guidance in China requires all multi-storey buildings to use stair evacuation, since ordinary lifts are not usually designed for use in emergencies. For buildings taller than 250m, it calls for evacuation measures to be reviewed by an expert panel. There is guidance in the USA that calls for the use of lift evacuation in buildings more than 128m, but this goes no closer to considering the case of super high-rise buildings, such as the 492m tall Shanghai World Financial Centre (SWFC). Stair evacuation of this 101-storey building would take two hours to evacuate and even then, the time could be much longer due to the effects of injury from trips and falls from thousands of people walking down hundreds of metres of stairs.

Fire evacuation strategy

Taking two hours to evacuate a whole building was considered too long for the SWFC's client, Mori Building Co., which commissioned Arup in 2003 to explore an alternative fire evacuation strategy, which would be quicker, safer and more considerate to the needs of its occupants.

SWFC includes a ground floor shopping mall, offices and a hotel plus a public observatory occupying the uppermost levels (Figure 4c1).

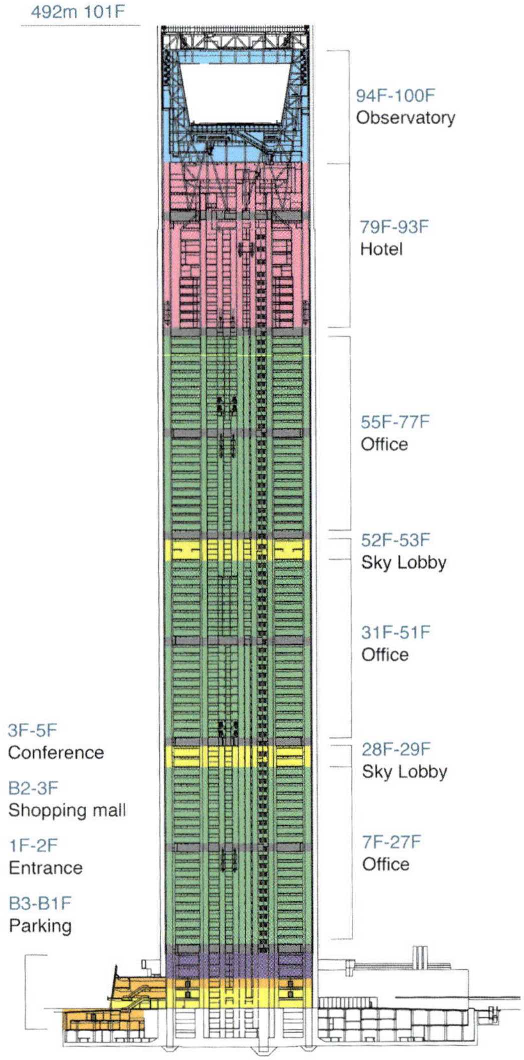

4c1
SWFC: a mixed-use building with areas open to the public
© Arup

Using recommendations from the US National Institute of Standards and Technology's investigations into 9/11 and using its own computer modelling techniques, Arup developed a fire safety strategy for the SWFC to incorporate lift-assisted evacuation.

Lift-assisted evacuation

Normal lifts are not generally used to evacuate people from a building because a lift shaft with openings on every floor poses greater risks to the spread of fire and smoke. Also, the lift equipment could be rendered useless in an emergency due to power failure from water damage. There is also concern that lift lobbies are not designed to accommodate large crowds of people evacuating a building and space restrictions could cause further problems, raising the anxiety levels of evacuees.

However, if these issues could be addressed, then the advantages of lift-assisted evacuation compared to stair travel could be realised – these advantages include faster and more efficient evacuation and the potential for fewer injuries en route.

When Arup developed the SWFC's fire evacuation strategy in 2003, lift-assisted evacuation had only been implemented in a handful of tall buildings across the world and none of these included public occupancy. By having different types of facility open to the public in the building, Arup had to consider the broadest range of occupants – infants, pregnant women, the elderly and the disabled – and their varying mobility challenges. According to the Hong Kong government's Census and Statistics Department, there has been a rise in the percentage of the population with disabilities from 5.2% in 2007 to 8.1% in 2013. The rise is attributed to an aging population and points to a trend which is likely to continue with time. Considering also that there are multiple flights of stairs in a tall building, these percentages could be much higher as moderately able people might also struggle to descend safely.

Refuge floors

SWFC's evacuation strategy involves activation of an emergency alarm and communication system to alert occupants that they should leave the building. These occupants evacuate via stairs to refuge floors before continuing down to street level via shuttle lifts or more stairs. Normal lifts, which serve every floor, would not be used in the evacuation plan and firefighters would use designated lifts to access the building.

Refuge floors, which are essentially unused and empty floors, provide a large, safe place for evacuees to rest or await rescue, as well as providing an opportunity to change the mode of travel to ground level (Figure 4c2). This enables the less physically able to change to lift evacuation and gives the option for those more able to continue by stair.

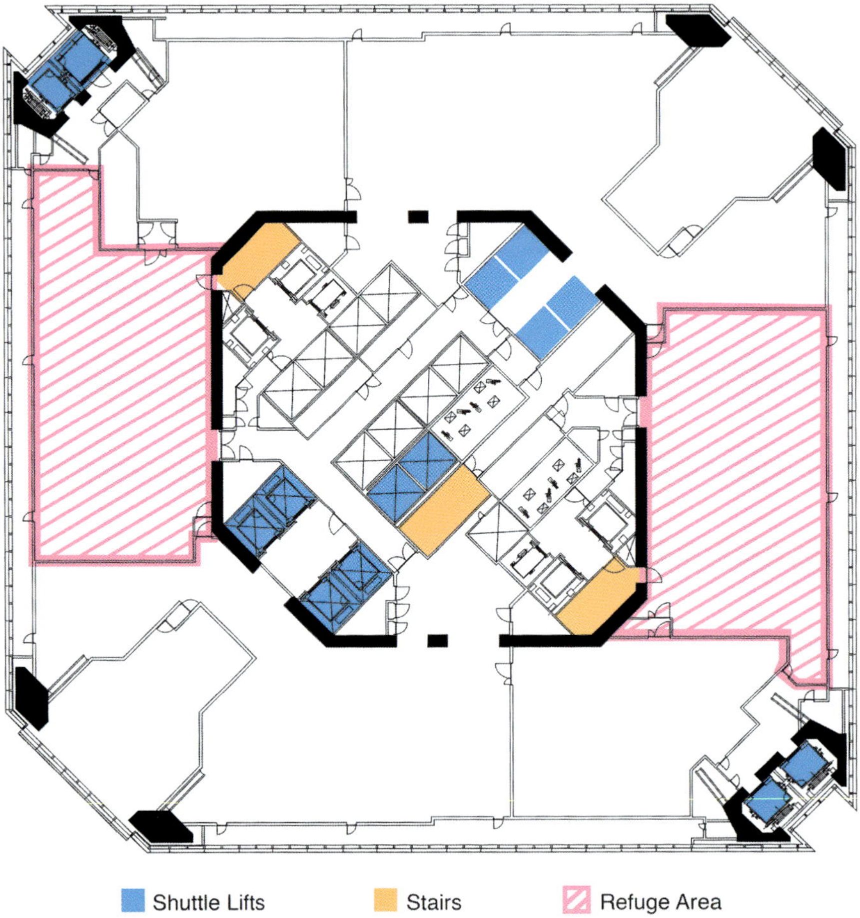

4c2
A typical refuge floor
© Arup

Under normal conditions, the shuttle lifts travel between the transfer sky lobbies at the top of the building and the ground floor. During an emergency, they are switched to evacuation mode where each lift serves one of the refuge floors in the building and the ground floor (Figure 4c3).

By housing the shuttle lifts in independent fire-rated cores, there is also a reduced likelihood of the lift shafts spreading smoke and fire since they do not open on normal floors where a fire may have started. Probability of water damage to lifts is also reduced.

Whole building evacuation

Arup also had to consider a worst-case scenario for whole building evacuation, despite maximum daytime office occupancy rarely coinciding with maximum occupancy in the hotel, mall or other areas of the building. A conservative 21,000 occupancy was considered in computer modelled evacuation simulations, with

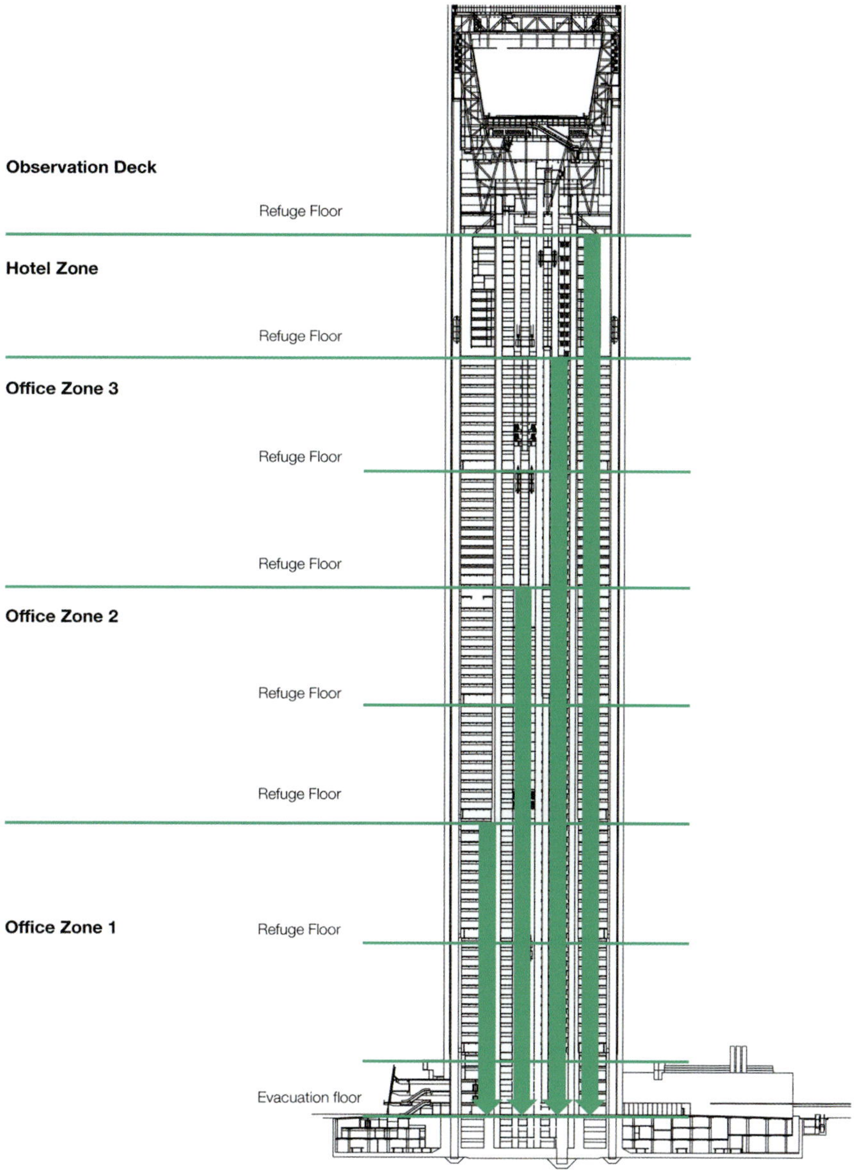

4c3
Shuttle lifts in evacuation mode
© Arup

further consideration given to modelling the response of mobility-impaired occupants (Figure 4c4).

The SWFC was one of the first tall buildings in China to consider using lift-assisted evacuation for public use. Arup's solution – to incorporate 13 shuttle lifts and refuge levels every 12 floors – reduced the time for 21,000 occupants to evacuate the building from 110 minutes to just 70 (Figure 4c6).

To ensure the fire evacuation plan would operate as expected, Arup produced a fire safety manual, which would need to be maintained and kept relevant throughout the lifetime of the building, and prescribed the need for:

- Regular fire safety training for the building manager to carry out phased and total evacuation drills
- Training for lift controllers on how to manage different emergency situations, including crowd control and effective communication
- Sufficient signage to affirm that evacuees can use lifts for evacuation
- Training of fire wardens on lift evacuation floors to facilitate evacuation.

Arup's innovative strategy for the SWFC has been highly regarded by the regional government, which shows more potential for its implementation in future tall building developments. The strategy is also relevant to underground structures, such as deep metro stations. The concept has already been applied to one of the new Mass Transit Railway (MTR) lines in Hong Kong.

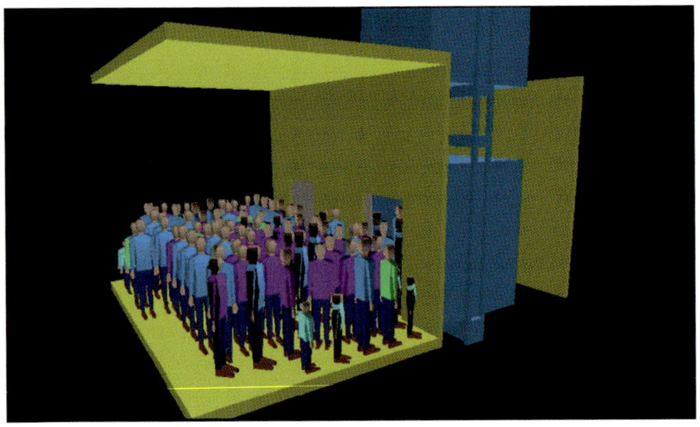

4c4
Close-up of part of the evacuation simulation showing how people queue at a refuge floor
© Arup

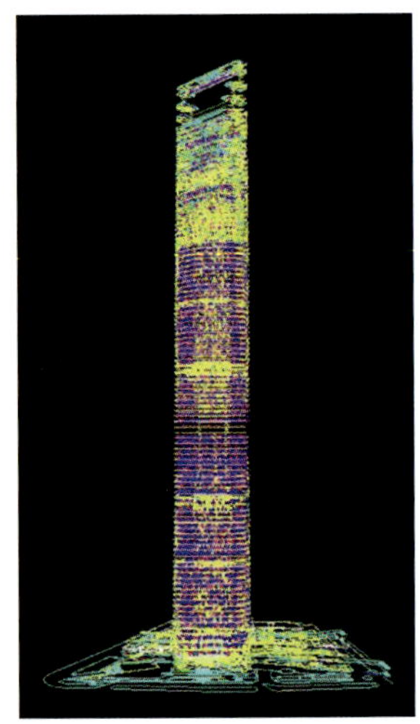

4c5
The simulation helps to estimate the time required to evacuate the whole building
© Arup

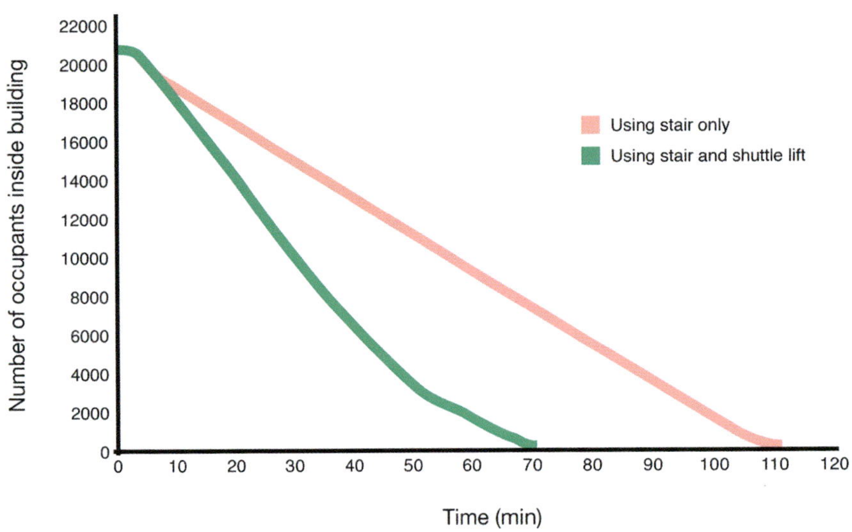

4c6
Reduced time taken to leave the building using lift-assisted evacuation
© Arup

4d

Case study 4d

China Resources Headquarters
Shenzhen, China

CONTRIBUTORS: ALEX TO, HAI LIN + QIAN ZHANG

Client:	China Resources Shenzhen Bay Development Co. Ltd
Design architect:	Kohn Pedersen Fox Associates (KPF)
Arup's role:	Structural, wind, geotechnical and fire engineer, façade consultant
Year of completion:	2018
Height:	392.5m
Number of storeys:	67
Gross floor area:	210,000m^2
Use:	Office
Main contractor:	China Construction Third Engineering Bureau

Sleek as a rocket about to launch into orbit, China Resources' headquarters building makes a bold statement on the Shenzhen skyline. Located in a region beset by typhoons, this landmark office building's height, shape, spire and external features have all been fine-tuned to improve its behaviour under different wind loads. By involving structural and specialist wind engineers early on in the project, much consideration has also been given to improving occupant comfort in high winds.

Architect KPF originally put forward three massing options for the tower at scheme design stage, each with a different cross-sectional shape – circular, square or triangular – and different architectural topping (Figure 4d1). For aesthetic reasons, the circular scheme emerged as the favourite early on, distinguished by its very slender perimeter columns that protrude the external glazing line.

To develop a deeper understanding of possible wind loading patterns and their effect on the structure, Arup proposed that these three options plus 11 others of varying height and external features should be examined in a special wind tunnel test. Its results would indicate which aerodynamic attributes should be carried forward in the design.

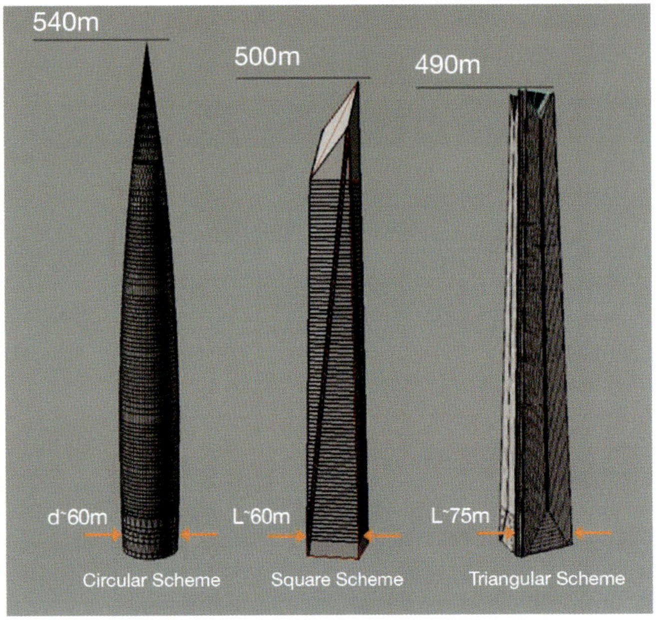

4d1
The three options for the China Resources Headquarters at schematic design stage
© Adapted from KPF

Special wind tunnel tests for aerodynamic optimisation

A model of the surrounding terrain and nearby buildings was made to a scale of 1:500 for the wind tunnel test. Wind of different speeds from 360 degrees, as predicted by climate data, was scaled down and applied using fans to a circular 550m tall baseline model. This initial test revealed that wind with incident angles of 80, 90, 100 and 110 degrees created the most adverse conditions. Winds from these angles only were applied in subsequent tests. Instrumentation at the base of each rigid building prototype allowed wind forces to be measured. These were then converted to "floor-by-floor forces" for detailed design calculations. Both cross-wind and along-wind responses were examined.

Wind tunnel test results confirmed that, compared to other prototypes, the circular tower experienced lower static effects and far reduced cross-wind effects. It also experienced the lowest cross- and along-wind moments overall (Figure 4d4). The results also revealed that peak dynamic moment increased with tower height – by as much as 10% for every 50m.

Results from the tests allowed architectural features to be scrutinised more closely for cost, structural efficiency, safety and ease of construction. Studying the behaviour of each model convinced the project team to adopt an enclosed, circular-based spire.

The tests also proved that features such as fins or slots had negligible effect on wind resilience, so were excluded in the final scheme. Such insights meant that the design could develop in a more certain and progressive manner from this point on, allowing a tight design and construction schedule to be maintained.

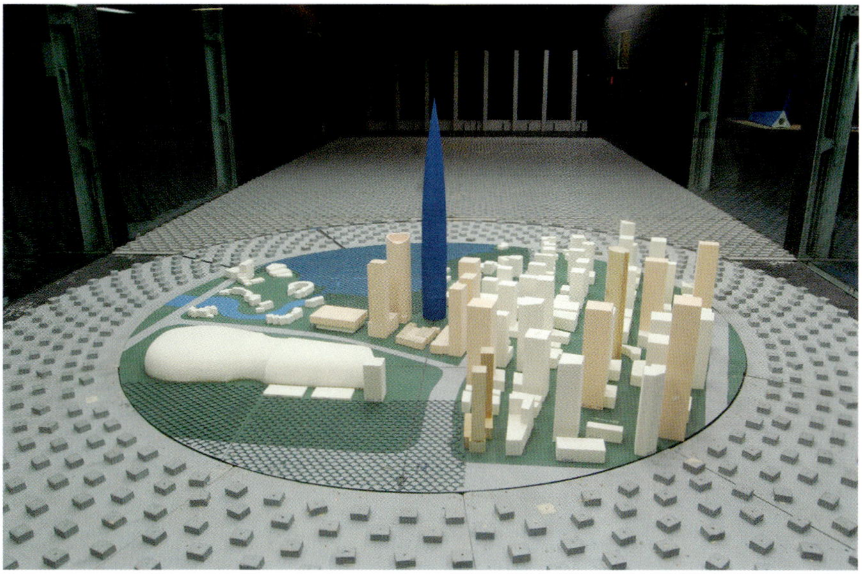

4d2
Baseline tower and proximity models in the wind tunnel test
© BMT

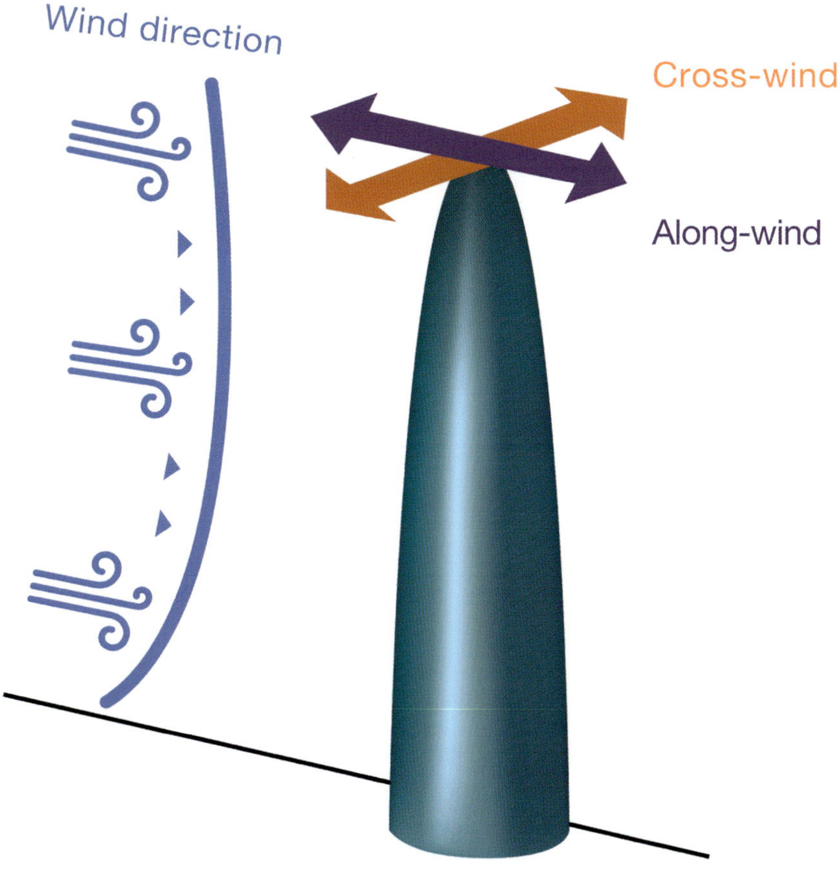

4d3
Wind-induced cross-wind and along-wind building responses were examined
© Arup

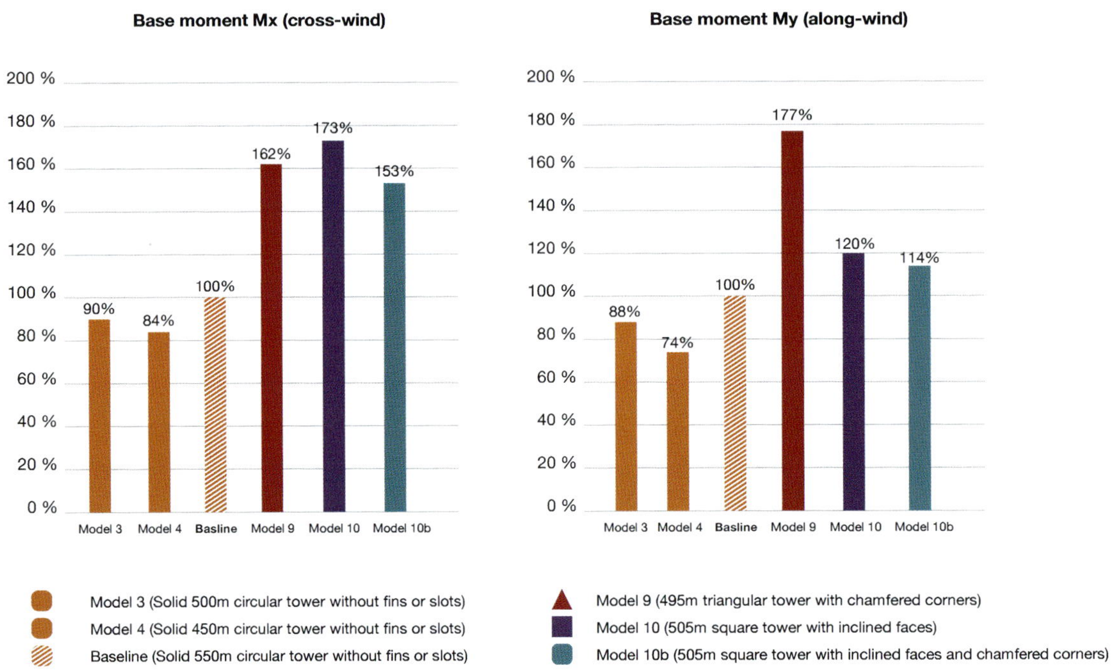

4d4
Comparison of peak along-wind and cross-wind base moments for six tower options
© Arup (adapted from BMT)

Through the course of the design and as more detailed information became known, the tower height was eventually fixed at 392m due to aviation constraints and included a 68m tall spire. Further wind tunnel tests carried out at a more detailed design stage confirmed the tower's dynamic design parameters, including its "peak acceleration".

Occupant comfort

Wind-induced acceleration occurs when a wind-loading pattern causes a building to oscillate or vibrate. Although it poses minimal risk to the structure, the movement can be disconcerting for occupants. Peak acceleration, experienced at the topmost occupied floor of a tower, is the common indicator for occupant comfort.

Early wind tunnel testing identified that the building's peak acceleration satisfied code requirements for one-year return period winds (which is 10 milli-g under the International Organization for Standardization (ISO) standard) but exceeded the limit for 10-year return period winds (which is 25–28 milli-g in Chinese code). To satisfy the latter, structural engineers would need to enhance the lateral stiffness of the structure or add a damping system.

Experience is everything

Although the building's peak acceleration value determined from wind tunnel tests after the design had been developed further deemed it acceptable according to Chinese design codes, Arup and the client felt closer scrutiny was still needed to understand the level of acceleration that would be most suitable for this building and its occupants.

4d5
The project team experienced several vibration modes in the motion simulator
© Arup

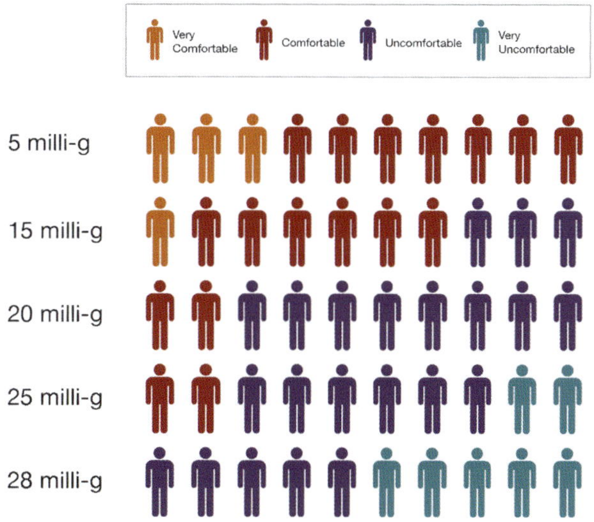

4d6
How people perceived wind-induced vibration at different levels of acceleration in the China Resources Headquarters motion simulator test
© Arup

A motion simulation experiment was commissioned to provide first-hand experience of acceleration. The simulator consisted of a motor which caused the floors and walls of a test room to vibrate at specific design frequencies and accelerations. This would replicate wind-induced acceleration on the topmost occupied floor of the tower. Ten people, including the head of the client's design team, participated in the experiment.

Feedback revealed that 80% of people felt noticeable movement due to the motion created by 25 milli-g of acceleration. When acceleration had reduced to 15 milli-g, only 30% felt the movement (Figure 4d6).

This evidence gave confidence to designers and the client that the building should be designed for an acceleration of 15 milli-g to give occupants an enhanced level of comfort. This is equivalent to the limit for residential buildings.

Damping solution

Since construction was already under way when the motion simulation's results became known, it was too late to stiffen the building using thicker columns or walls without profoundly affecting the building's architecture. This left supplementary damping as the only other stiffening option to reduce the building's peak acceleration. Either a tuned mass damper (TMD) at the top of the tower or a series of viscous dampers located on plant floors would be most appropriate. However, the high cost of TMDs ruled this option out, leaving viscous damping as the most viable solution.

An integrated viscous damping system was chosen, consisting of eight outrigger dampers across plant and refuge levels 47 to 49. These took the form of stiff outrigger "arms" which connected the reinforced concrete core walls to the perimeter steel columns. Customised viscous dampers were positioned between the outrigger and perimeter column. To maximise the damping effect, structural engineers optimised the damping system's layout and its response to wind from 360 degrees.

Another concern revolved around the tower's slender perimeter columns and whether they could support the large forces induced by the outrigger mechanism under the 10-year return period wind load. This load capacity became one of the limiting factors in the damper system's design.

Integrating dampers in this way significantly increases a building's ability to control movement due to wind, while also reducing lateral acceleration. This has benefits for the structure, which can be designed for lower wind forces.

This is the first time high-powered viscous dampers have been used to control wind vibration in mainland China.

China Resources Headquarters has been designed, not just to code requirements, but to more advanced performance characteristics that have been verified through physical testing. The project highlights the importance of wind design in tall buildings and how, when considered early on, this allows a building's form to develop more holistically, leading to a more efficient design.

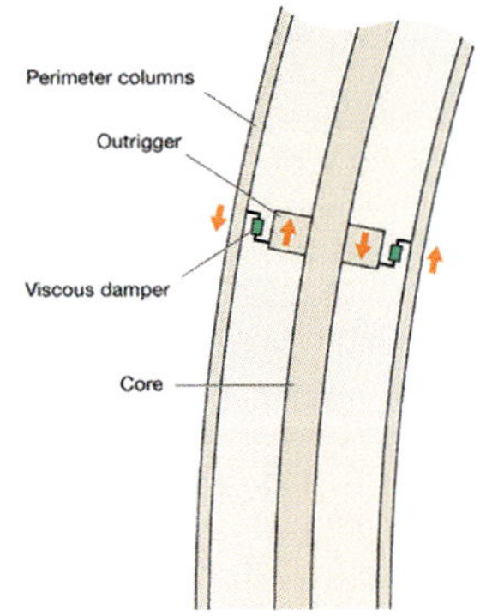

4d7
Concept of the outrigger damper
© Nigel Whale/Arup

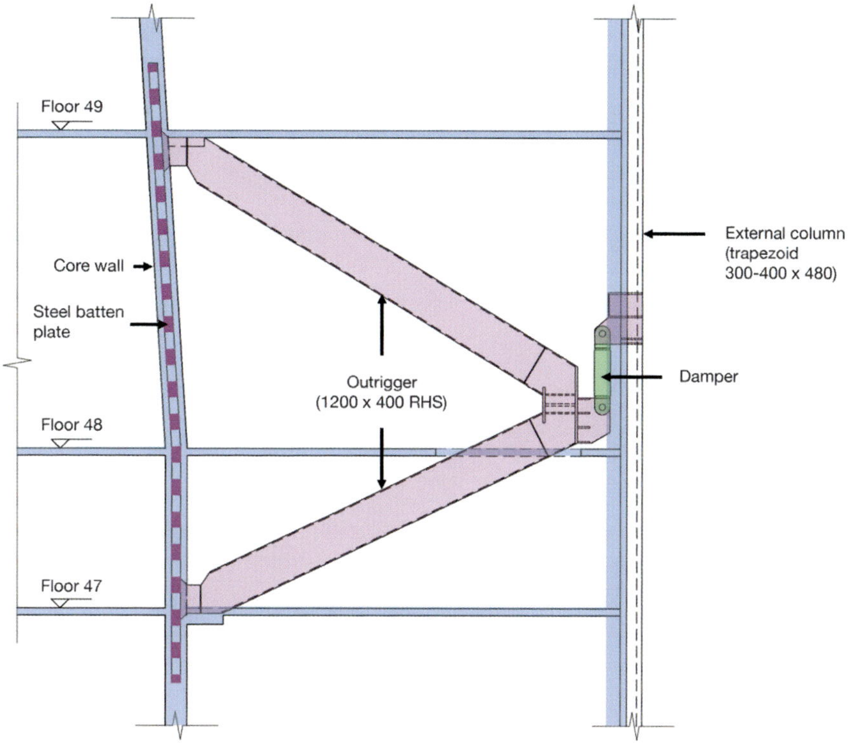

4d8
Viscous outrigger dampers located at a height reduce wind-induced vibration and create a more comfortable environment for occupants
© Arup

4d9

China Resources Headquarters will be the tallest building in the Houhai area of Nanshan when it opens in 2018
© China Construction Third Engineering Bureau

Horizontal load-resistant structural system

The main structure is made up of a "tube" of 56 closely spaced slender steel trapezoid columns and a reinforced concrete core. Compared to tall buildings with mega column structural frames with only a sparse arrangement of columns, this system is quicker and more cost-effective to build. The system also facilitates a column-free interior, while still possessing the stiffness required for earthquake design. Lateral forces are mainly resisted by the core, since the perimeter steel tube resists only a very limited proportion. It is the first time this structural system has been applied to a building of this height in a seismic area in mainland China.

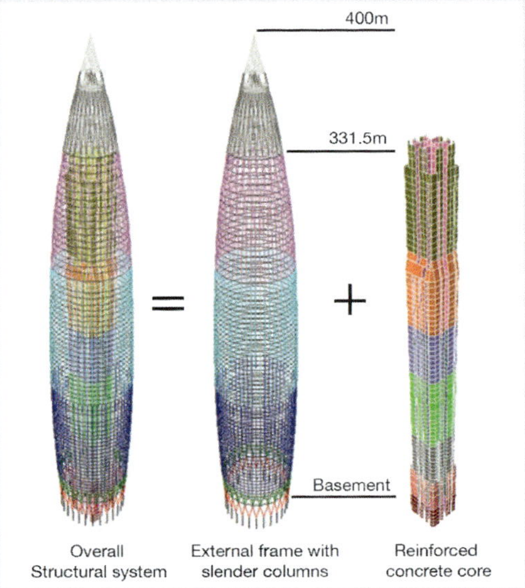

4d10

Structural system of China Resources Headquarters
© Arup

Chapter 5

Green building

Tall buildings are a major source of carbon emissions due to the vast quantities of energy required to heat, cool and ventilate them. They can also have an adverse effect on their immediate surroundings by blocking sunlight and airflow and raising local temperatures.

The survival of tall buildings in our urban landscapes depends on their ability to consume less energy and to function more harmoniously with neighbours, as well as be adaptable for climate change.

Technological advances in active and passive sustainable design mean that buildings of today can benefit from more energy efficient, low carbon systems than their predecessors. This could be by automatically adjusting artificial lighting to complement natural levels or by encouraging fresh air ventilation when pollution levels and external temperatures allow. Many of these systems can be retrofitted, making it cost-effective to refurbish an existing building with the latest, low carbon equipment.

Innovatively designed new-builds now have façades with integrated passive attributes and cooling strategies to counter solar gain – a major concern for building tall and green; and where an entire urban district is already suffering from the effects of high-density development and pollution, there is also a green answer: vegetation on walls and roofs to reduce the tendency for new additions to absorb and radiate heat. Sizeable openings in new buildings can also encourage cross-ventilation so that warm, polluted air can disperse.

5a

© Marcel Lam Photography

Case study 5a

China Resources Building
Hong Kong

CONTRIBUTORS: VINCENT CHENG + GARY CHEUNG

Client:	China Resources Property Ltd
Architect:	Ronald Lu & Partners (Hong Kong) Ltd
Arup's role:	Building sustainability, LEED and transport planning consultant
Year of completion:	1983 (renovation completed in 2013)
Height:	178m
Number of storeys:	50
Gross floor area:	99,000m^2
Use:	Office
Main contractor:	CR Construction Co. Ltd

Concerned that tenants of the 25-year-old China Resources Building (CRB) in Hong Kong would leave for more modern offices with lower running costs, the building's owner, in 2008, decided to refurbish the premises to make them more appealing and energy efficient.

The drive to refurbish rather than demolish and rebuild came from the client's desire to differentiate the offices from others as the more sustainable and forward-thinking option – with the hope that it would also attract like-minded businesses to rent space in the building.

The greener option

Refurbishing the 50-storey office block also made sense commercially since it would be cheaper to reuse the existing structure and tenants could remain in situ. There were also multiple environmental advantages, since reusing the structure avoided the carbon emissions associated with demolishing it, transporting waste away and rebuilding it using energy intensive materials such as concrete and steel.

With about 30% of grade A office buildings in Hong Kong built before 1990 (Figure 5a2) Arup also realised that there was great potential for more buildings to be refurbished to make them more commercially attractive and sustainable.

5a1
The China Resources Building before refurbishment
© Arup

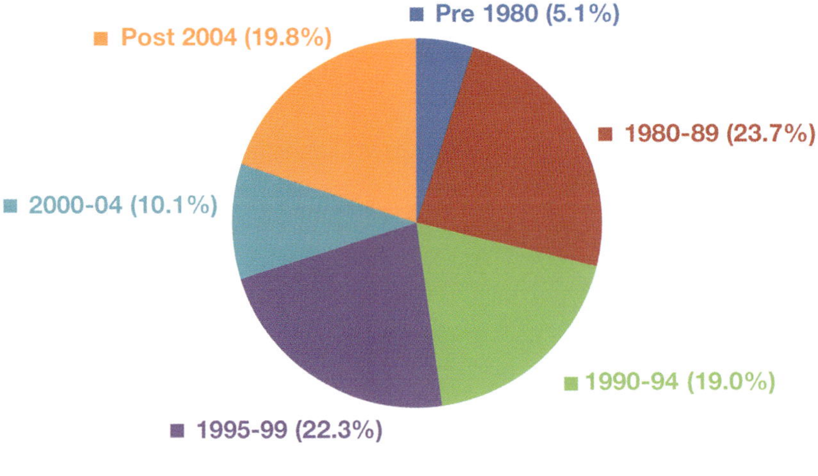

5a2
Hong Kong grade A offices – stock distribution by age
© Arup (based on Hong Kong Property Review 2015 by Hong Kong Government Rating and Valuations Department)

Ahead of working on the CRB project, Arup established a "Five-step Strategy", which could be applied to any outdated building to determine its potential for sustainable refurbishment. The steps include defining the baseline energy consumption for a building before determining a range of improvements to achieve the desired sustainable outcomes. Consideration was also given to improving existing management and maintenance systems, and forecasting the building's resilience to climate change.

Energy audit

Carrying out an initial energy consumption and environmental performance audit pointed to where improvements could be made in the CRB. Published manufacturer's data, meter readings and information from utility bills were used to establish the baseline for the building, from which Arup was able to identify that lighting, water-efficiency and HVAC (heating, ventilation and air conditioning) systems could be vastly improved to meet higher environmental and efficiency standards. Although time-consuming (taking approximately one month to complete) the information collected was essential to demonstrate the advantages and feasibility of refurbishing.

Setting the scope and targets

Some improvement options had to be ruled out due to building constraints. For example, the prospect of adding a green roof was ruled out due to loading restrictions and, since the building occupied most of the site's footprint, there was very little scope for adding landscaped areas. Both of these features would have improved the environmental credentials of the building.

Harvesting rainwater for reuse was also not practical due to lack of space to store it in the basement. Since seawater rather than potable water is already used for flushing toilets in Hong Kong, rainwater harvesting would not yield a significant environmental or capital return. Installation of more water-efficient fixtures in washrooms and kitchens was considered a better investment.

It was considered very important for the building's energy and environmental performance to undergo independent assessment, so that the building owner could promote the sustainable credentials of the offices to existing and potential tenants. Arup proposed that the owner follow the US Green Building Council's Leadership in Energy and Environmental Design (LEED) framework, which is internationally recognised.

The refurbishment strategy was then developed according to the requirements of the highest possible standard that the CRB could achieve under the LEED framework. Inherent site constraints and available technology limited the potential for achieving a higher standard than "gold". For example, the market for materials containing recycled content or being from sustainable sources was not as developed in 2008 as it is now. At the time, these materials would also take too long to import and involve transporting them long distances, nullifying their sustainability credentials.

Limiting disruption to tenants

Arup realised early on that to maximise the benefits of more energy-efficient HVAC systems, a replacement curtain wall façade would be beneficial. This would be more airtight and better at transmitting light into the building while limiting the amount of heat transferred. But installing a new façade using conventional means (bamboo scaffold and netting) would shroud the building, creating an undesirably dark environment for tenants. There would also most likely be significant noise and dust associated with carrying out the work. Bearing in mind that a client's greatest concern when it comes to refurbishing is to limit the disruption caused to existing tenants, the method of retrofitting had to be discreet.

Working closely with the contractor, a solution using elevated platforms was put forward where the new façade would be fixed onto the building first, and only after completion would the original glazing be removed. This would reduce noise disturbance to office workers as the original windows would act as a buffer to external works. It also meant that views out of the windows and, hence, natural light levels, could be maintained.

5a3
An electrical elevated platform was used to allow flexible façade installation
© 2017 China Resources Property Ltd

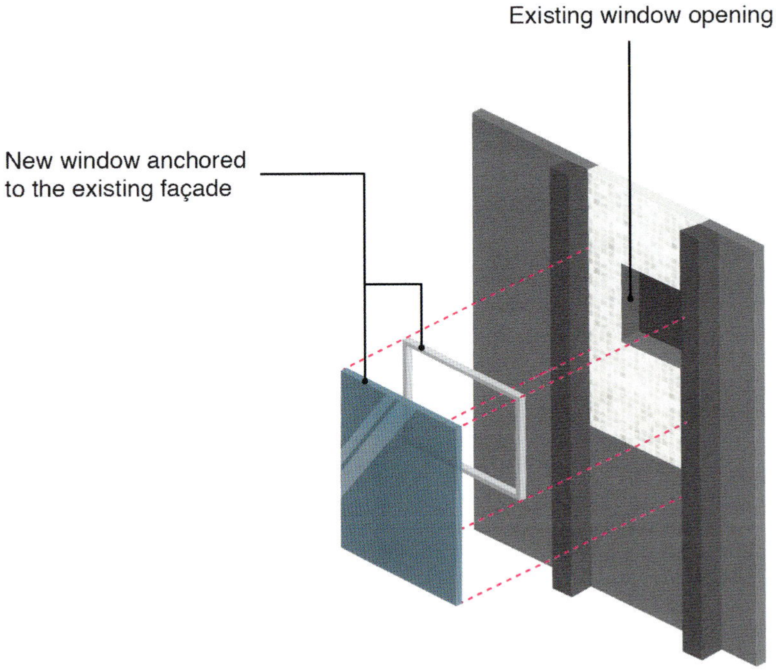

5a4
New window simply anchored to the existing façade to minimise disturbance to the tenants
© Arup

Low-energy lighting strategy

5a5
High-performance building façade with low-e glass panels to minimise the amount of UV and infrared light entering the space, while allowing visible light through
© Arup

Inclusion of daylight sensors to automatically dim artificial lighting on bright days offers energy savings. Installation of occupancy sensors also ensures lights are switched off when no one is in the office.

To maintain the CRB's profile on the Hong Kong skyline, the client was keen to illuminate the façade. As this would consume a lot of energy (in 2008, LED lighting was not as efficient as it is now) Arup devised an LED lighting strategy where only the top of the building would be lit up at night, with changing colours to make the building stand out.

5a6
Exterior lighting of CRB
© Arup

Fit for the future

The CRB was the first major refurbishment project in Hong Kong to receive a Gold LEED rating (core and shell) in 2012. The improvements made include a 27% water saving, equivalent to saving 5,650m^3 of water per year. It has also achieved 8.3% reduction in greenhouse gas emissions, equivalent to a reduction of 433 tons of carbon dioxide. Energy consumption has also reduced by 8.3% or 550 MWh, the equivalent of 1,700 fluorescent light tubes running 24 hours a day, for a year.

Arup also assessed the building's performance against a predicted temperature rise due to global warming. Its analysis revealed that the refurbished building would continue to use less energy than its predecessor for 40 years, despite the building requiring more cooling over this period due to global warming to maintain a high level of comfort. Perhaps most reassuring for the client, a survey carried out after refurbishment revealed that satisfaction had increased by 25% compared to before the works had taken place.

Other low energy and sustainable features:

- Complementing the more energy efficient HVAC system, the new low-e laminated glass panels reduce solar heat gain (only 5% of solar heat energy transmitted) without limiting the transmission of natural light into the building.
- To improve the flow of people through the building, more energy efficient elevators were recommended with reconfigured lift lobbies on the ground and second floor to reduce congestion and waiting times.
- Demand Control Ventilation (DCV) sensors detect the amount of carbon dioxide expelled by occupants on a floor and ventilation is automatically varied to suit demand.
- Low volatile compound containing fixtures and finishes were used to minimise harmful chemicals released into the air.
- Ninety-seven per cent of the CRB's existing envelope, structural core, roof and floors were retained in this refurbishment project. Removal of a plinth at ground floor level was the only major structural alteration. This was done to encourage more natural ventilation around lower parts of the building.
- Nearly 2,000 tonnes or 81% of waste from the refurbishment project was reused or recycled and new building materials with recycled content, or locally sourced (up to 800km away), were specified where practical.

5b
© Zhou Ruogu Architecture Photography

Case study 5b

Ping An International Finance Centre
Shenzhen, China

CONTRIBUTORS: VINCENT CHENG + WAI-HO LEUNG

Client:	China Ping An Group
Architect:	Kohn Pedersen Fox Associates
Arup's role:	Building sustainability and LEED consultant and fire engineer
Year of completion:	2016
Height:	599m
Number of storeys:	115
Gross floor area:	459,525m^2
Use:	Office
Main contractor:	China Construction First Group Construction & Development Co. Ltd

Standing head, neck and shoulders above other tall buildings in the Futian district of Shenzhen is the 599m tall Ping An International Finance Centre (Ping An IFC). An office block sitting on a 10-storey retail podium, the building accommodates around 15,656 workers on a footprint of just 3,800m^2. Servicing such a tall building and keeping its population comfortable uses vast amounts of energy. Realising this, engineers were challenged by the client to design the building to be the most environmentally responsible super tower in the world.

Highly green

Using the US Green Building Council's Leadership in Energy & Environmental Design (LEED) framework, energy saving and efficiency measures undertaken at Ping An IFC have resulted in 18% energy cost savings (compared to the Energy Standard for Buildings except Low-rise Residential Building published by American Society of Heating, Refrigerating and Air-Conditioning Engineers – ASHRAE Standard 90.1(2004)). The building has also been granted a pre-certification LEED Gold Award and is expected to be one of the first 500m+ tall buildings in the world to achieve this level of environmental recognition.

For the building to qualify for these accolades, engineers had to first address two fundamental challenges associated with super-tall buildings: transporting large numbers of workers to their destinations efficiently and

5b1
Ping An IFC is much taller than neighbouring buildings, so is exposed to high levels of solar gain for the majority of its height
© Zhou Ruogu Architecture Photography

creating a comfortable office environment sustainably. By selecting energy efficient double-deck lifts and creating zonal destinations for workers, engineers cracked the vertical transportation problem early on in the design process.

Dealing with occupant comfort was a challenge which had to be tackled on many fronts. Because the majority of the tower is above the Shenzhen skyline, heat from the sun penetrates the building, requiring large areas to be cooled and ventilated artificially. Energy efficient equipment was specified, complemented by other passive solutions developed by Arup, and the result is a total energy saving of 5,670MWh per year as compared to the ASHRAE baseline. One of these passive solutions is an innovative integrated façade with multiple passive attributes together with energy efficient cooling strategies to counter the heat gain effect.

Energy efficient, integrated façade

Integrated glazing units with a low-emissivity coating were specified for the building. They allow light to penetrate the façade, but limited heat transfer. The high performance glazing maximises transmission of light into the building to reduce the dependency on artificial lighting. By optimising the portion of vision glass and spandrel area on the façade envelope, natural light is able to penetrate deeper into office floors.

A direct addressable lighting interface system with wireless control was also installed; allowing two-way communication between the ballasts and the controllers, the lighting system automatically adjusts the level of lighting efficacy on a floor by taking into account natural daylight levels. A bonus of the wireless facility is that the sensing receivers are the major part of a building-wide system, allowing all tenants to benefit equally without the further wiring required for re-zoning. Occupancy sensors also detect whether a floor is occupied and automatically switch off if the office is empty. Both features reduce the amount of energy consumed.

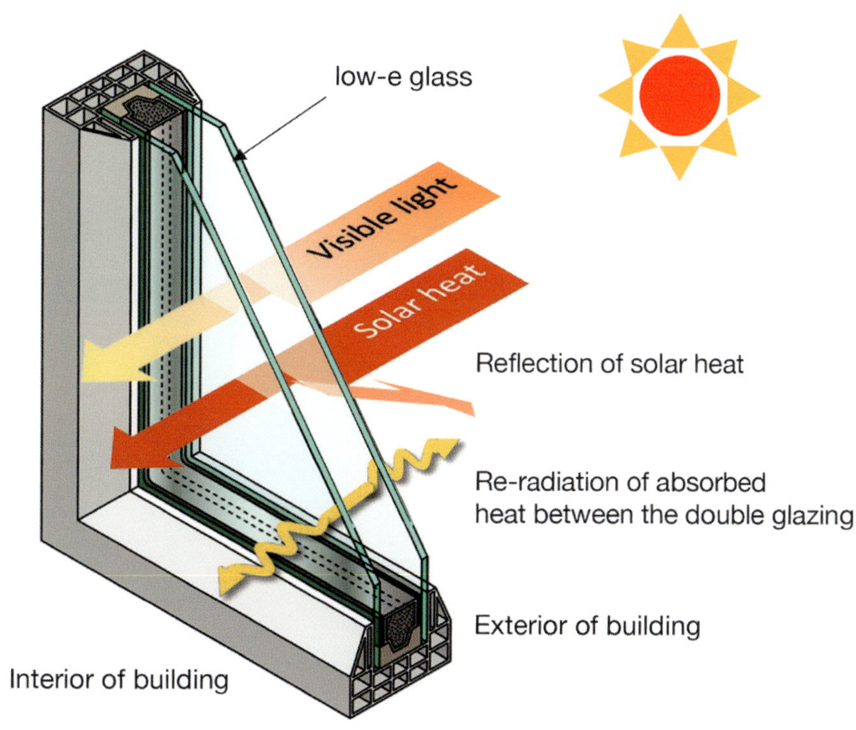

5b2
Integrated glazing units allow light to penetrate the office floors with a limited amount of heat transfer
© Shutterstock/Zern Liew/Arup

Free cooling

A rise in building height correlates to a decrease in outside air temperature. The phenomenon is significant in super high-rise buildings such as the Ping An IFC where the temperature difference between the ground and top of the building is nearly 3.5°C. Arup carried out computational fluid dynamics analyses to understand the outside air pressure acting on the building envelope in order to make best use of this cooler air. In specific locations, and taking into account different wind pressures, cool, fresh air is allowed to enter the building, in a process known as "free cooling".

The free cooling system allows air to be drawn in from mechanical floors at the top of the building and feed into the building's primary air unit (PAU). The cooler outside air (during winter months and at certain times of day) then transfers from the PAU to each floor by mechanical fan and vertical ducts within the core of the tower which reduces the reliance on artificial cooling. Collecting outside air from higher levels of the building with a lower air temperature extends the length of time free cooling is available. To avoid the problem of positive wind pressure pressing on louvres and obstructing air intake, a buffer zone was created between the external louvre and internal air duct intake.

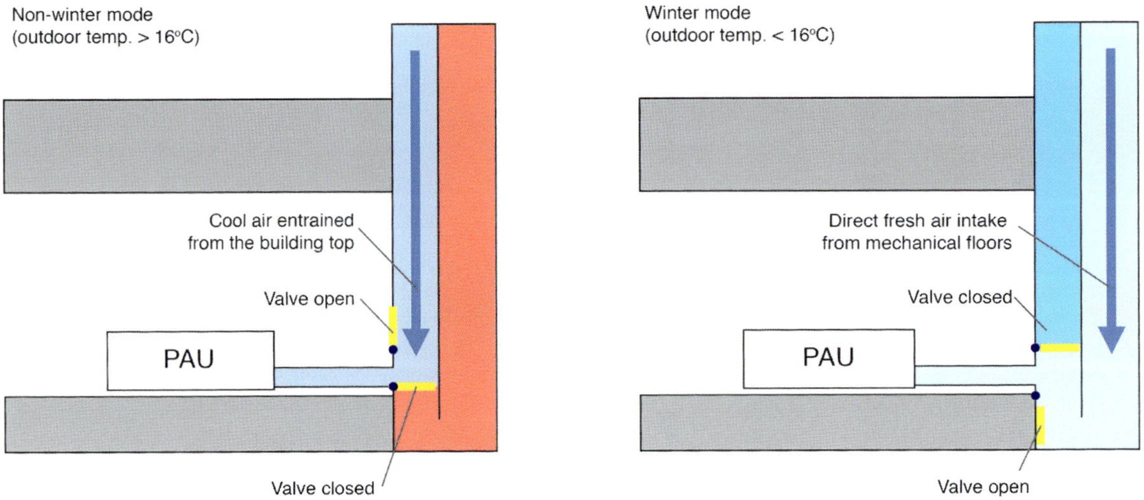

5b3
The lower outside air temperature higher up the building is exploited to support the building's cooling strategy
© Arup

Combating solar heat gain

Arup also carried out a sun-path study to track where and when the building would be most susceptible to heat gain. This identified the orientation, position, size and shape of protruding architectural features that could offer shading. The study led to the incorporation of external vertical stone fins running the full height of the building to block the sun's rays. Fin spacing was based on results from the solar tracking analysis, while the overall form was developed in conjunction with the architect to suit the building's aesthetic profile. As a consequence, the façade

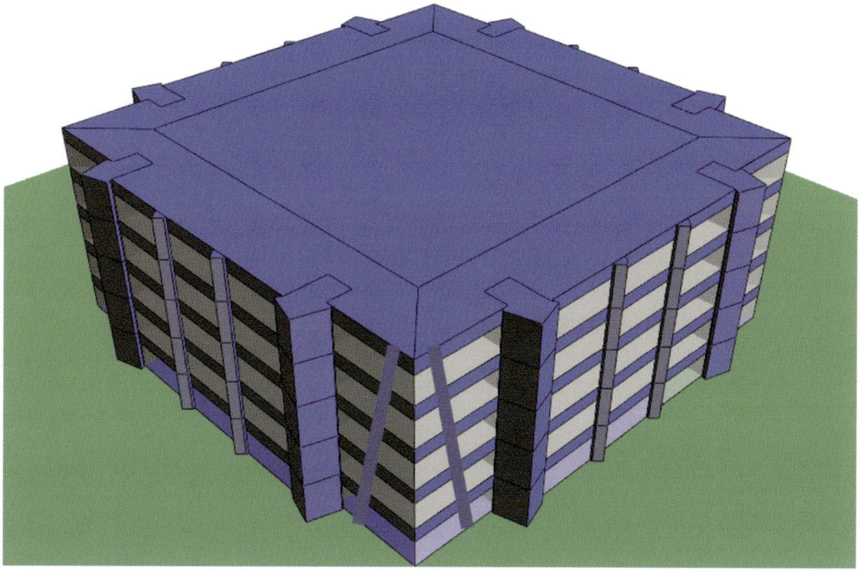

5b4
The effect of fixed shading provided by architectural features on the building was explored using a computer model
© Arup

5b5
The location, size and spacing of stone fins running the full height of the building was based on the results of thermal analysis and a solar tracking study
© Zhou Ruogu Architecture Photography

performs 20% better than would be expected from local design codes, leading to a reduction in space cooling requirements and, hence, energy consumption.

To reduce aerial heat absorption, the roof of the tower is covered with solar reflecting materials and the podium roof has planted areas.

Ice storage and demand control ventilation

To complement the high efficiency air conditioning (AC) system adopted in Ping An IFC and to take advantage of lower night-time electricity tariffs in Shenzhen, ice is made overnight in the building's chiller units for AC supply during daytime periods. The ice is melted by day to cool the building and reduce operational energy costs. Further energy savings from the AC system are made by monitoring carbon dioxide levels in the building (from air expelled by occupants) to control the level of ventilation to suit levels of occupancy. This is known as demand control ventilation.

Travelling with less energy consumption

To transport people efficiently without consuming vast amounts of energy, a lift with a regenerative drive system was specified for the Ping An IFC. The system allows power to be regenerated under two conditions: when light loads are going up or when heavy loads are coming down. Elevator planning formed a significant part of the building's early design.

Sympathetic to surroundings

The project's sustainability merits are not purely limited to optimising energy use, but also consider the effect the building has on the local environment. Inclusion of open space and a set-back at ground floor enhances wind penetration to improve the surrounding area. Glazing material with lower reflectivity also reduces glare being a nuisance to neighbours.

Ping An IFC uses a range of cutting-edge technologies and strategies to make it greener. It serves as an example for all future super towers to reduce their impact on the environment and secure the sustainability of tall buildings for many years to come.

5b6
The design of the building minimises its impact on the surroundings
© Kohn Pedersen Fox Associates PC

5c
© Kenny Ip

Case study 5c

Hysan Place
Hong Kong

CONTRIBUTORS: VINCENT CHENG + WAI-HO LEUNG

Client:	Hysan Development Co. Ltd
Design architect:	Kohn Pedersen Fox Associates; Dennis Lau & Ng Chun Man Architects & Engineers (HK) Ltd
Arup's role:	Environmental and building sustainability consultant
Year of completion:	2012
Height:	204m
Number of storeys:	36
Gross floor area:	66,511m^2
Use:	Office and retail
Main contractor:	Gammon Construction Ltd

Hysan Place is a gentle giant of a building which treads lightly on Hong Kong's space-constrained Causeway Bay commercial district. The vibrant area is a victim of its own success where high-rise, high-density development has led to traffic congestion, a shortage of open space and poor air quality, all of which threaten the area's future prosperity.

Aware of these issues, developer Hysan Development Co. decided that when number 500 Hennessy Road in the heart of Causeway Bay was due to be demolished and rebuilt, the new building should go some way towards alleviating the environmental problems of the area.

The new building, Hysan Place, had to, without doubt, be functional, stylish and modern, but the developer also wanted sustainability to be at the core of its design. This put great emphasis on all parties to work collaboratively to realise the full sustainable potential of the building.

Urban windows

Hysan Place is a 36-storey building, which is made up of a lower retail podium up to level seven with a slimmer office tower on top. To maximise tenancy options, slotting next to the tower, and on top of the podium, is a flexibly designed

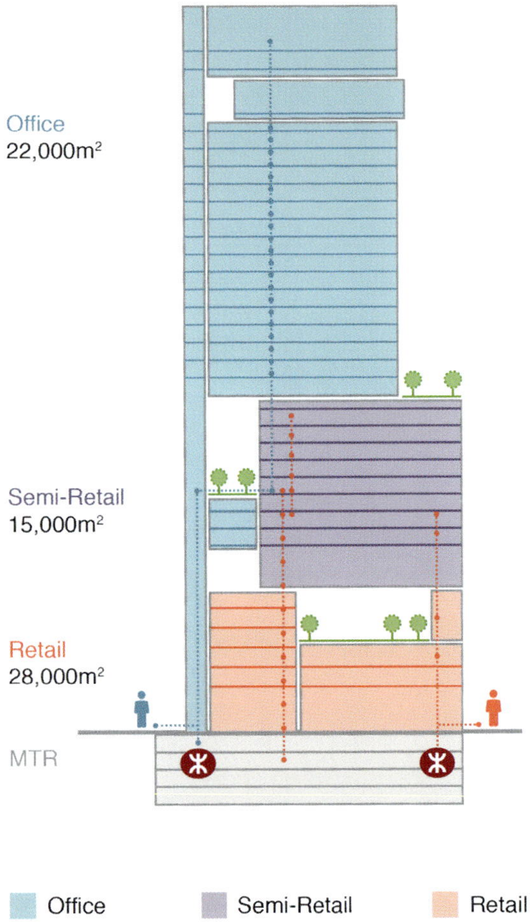

5c1
Hysan Place comprises an office tower, semi-retail block and retail podium
© Adapted from Kohn Pedersen Fox Associates

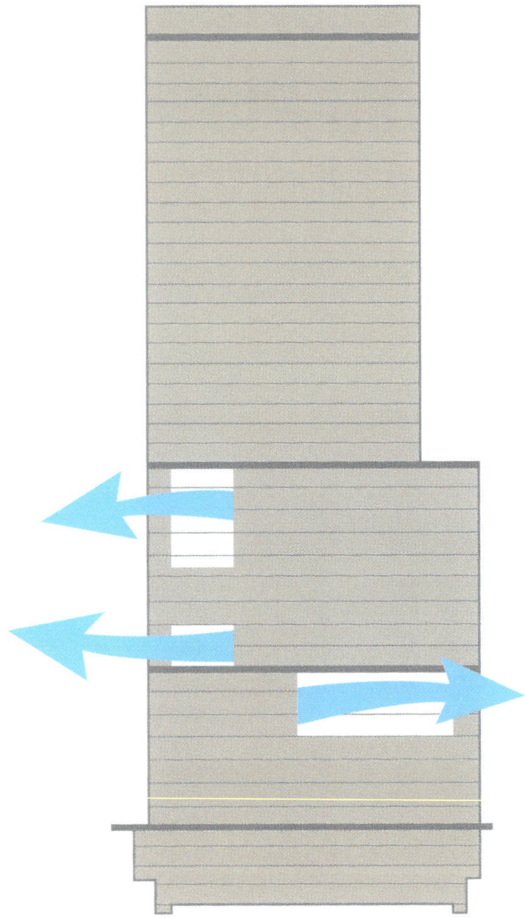

5c2
Urban windows facilitate natural ventilation to improve the microclimate of the surrounding area
© Adapted from Kohn Pedersen Fox Associates

nine-storey block, which can be adapted for retail or commercial use, as dictated by market needs.

"Urban windows" form the building's most distinguishable features, punching through the structure at three locations at low level (Figure 5c2). These openings allow air to circulate around the building to help counter the problem of stagnant air which blights the neighbouring built-up area.

This poor air quality is due to pollution caused by vehicle exhaust fumes and poor airflow around the densely packed buildings of Causeway Bay. As a consequence, the dirty air cannot disperse and creates a caustic environment for pedestrians, particularly in warm weather.

As sustainability consultant, Arup put forward the idea of including the urban windows, working closely with the architect and client to develop their

5c3
An air ventilation assessment was carried out to justify the effect of the urban windows on the wind velocity
© Arup

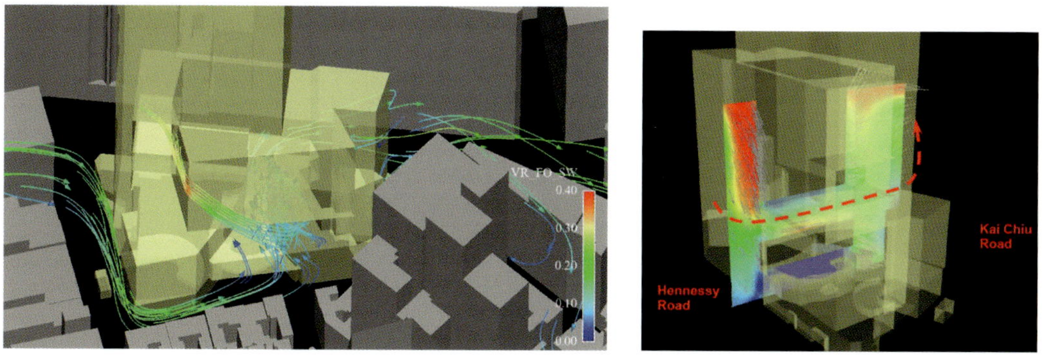

5c4
CFD analysis provides visual proof of how urban windows work
© Arup

unique attributes. An air ventilation assessment (AVA) was also carried out, as was mandatory at the time, to understand how the new building mass would affect the flow of wind at surrounding pedestrian level.

Taking into account climate data for prevailing wind directions, intensity and probability of occurrence, the appropriate location, number, angle and size of urban windows was developed and tested using computational fluid dynamics (CFD) analysis models. A further wind tunnel test provided even more insight into the positive effect these features would have at street level and on top of the podium.

Gardens in the sky

Exacerbating the problem of stagnant and polluted air at Causeway Bay is a phenomenon known as the "heat island" effect. This is where closely packed buildings absorb and store heat, later releasing it back into the environment and raising the local ambient temperature compared to further afield.

a

b

5c5

Vertical and horizontal planting reduces the building's ability to absorb, store and release heat, thus reducing its contribution to the "heat island" effect. (a) roof garden, (b) green wall, (c) Sky Garden

Images a–b © Arup

Image c © Cheung Tsun

c

Improving air circulation via urban windows allows heat to disperse more efficiently, but it is the vegetated areas on the roof, internal courtyard walls and the level 4, 7 and 16 Sky Gardens which have the greatest impact on reducing solar heat absorption and the heat island effect.

A total vegetated area of about 1,800m^2 has been achieved at Hysan Place, which equates to 40% of the site area. These areas of greenery also create a sanctuary for workers and shoppers using the building. Moreover, the roof garden functions as an urban farm where office workers can enjoy cultivating crops and schoolchildren can visit to learn about sustainable living. It provides a platform to share ideas on using natural resources and reducing energy consumption, spreading the message much wider into the community. A "wetland" on the sixteenth floor Sky Garden offers additional benefits by treating wastewater through its biofiltration system.

Elsewhere in the building, other integrated design features mean that the building itself operates more sustainably by consuming less energy and making use of natural resources.

Fresh air ventilation

Working closely with the architect and mechanical, electrical and public health (MEP) engineers, operable vents near the floor and ceiling are included within each level's façade system to allow fresh air ventilation.

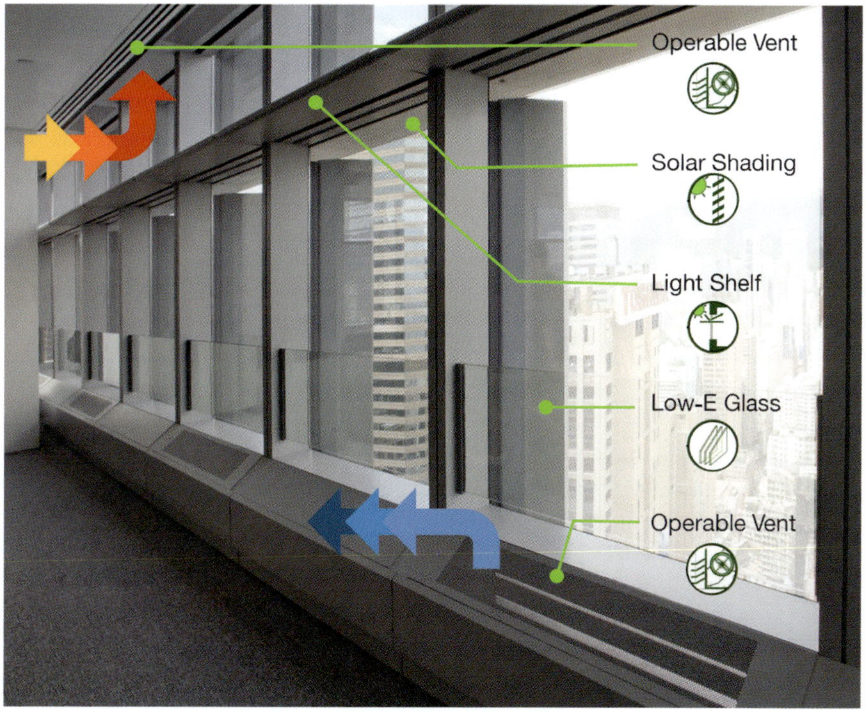

5c6
Hysan Place's high performance curtain wall system to create a low energy and comfortable environment for office workers
© Arup

Connected to a local weather station, the system is activated when pollution levels are low enough for natural ventilation to be used safely. During these periods, a green light alerts workers that the vents can be opened to allow fresh air into the office, easing the dependency on artificial ventilation, and so reducing energy consumption.

Arup reviewed annual weather data in the Causeway Bay area and its effect on outdoor air quality, pollution and humidity to determine the likely periods of time when natural ventilation could be utilised. This determined whether the feature would be cost-effective and yield the environmental returns desired. The location and size of ventilation louvres were designed based on this data and airflow results from CFD analyses.

Energy efficient façade

The curtain wall façade also includes many energy efficient features such as perimeter low-emissivity double glazing which limits heat energy penetration and maximises light. On the northern elevation, an internal "light shelf" reflects light deeper into the office and reduces the reliance on artificial lighting. Externally, horizontal and angled vertical fins on the western elevation also provide optimum shading where solar heat gain is greatest.

Core relocation

One of the most fundamental changes to the internal structure of Hysan Place to yield environmental benefits has been to locate the stair and lift core to the south of the building. Since the southern elevation experiences the most intense solar heat gains, a highly insulated core in this location helps block heat transmission into the building. It also has the effect of reducing energy consumption by not requiring as much cooling as a typical office space. Sunlight simulation demonstrated that there is little change in overall daylight available in the interior, despite light from the south being partially blocked by the core wall. This also suited the client's desire for a larger floorplate which could enjoy more fully the spectacular harbour views to the north.

Platinum LEED and BEAM Plus certificates

Specifying efficient water fixtures and maximising the use of recycled materials and low volatility compounds also contributed to the sustainability credentials of the building. Connection to underground train networks also reduced the reliance on travelling by car, so reducing traffic congestion and pollution in the area.

Designing the building to step back at street level to widen the gap between Hysan Place and neighbouring buildings also created more space for pedestrians, something of a rarity in Causeway Bay. Local planning regulations also permitted construction of a taller, but slimmer podium in exchange for more space at ground level to encourage these benefits.

The building has achieved the highest recognition under the internationally recognised US Green Building Council's Leadership in Energy and Environmental Design Core and Shell (LEED-CS) Version 2.0 programme, and the BEAM Plus Version 1.1 programme of the Hong Kong Green Building Council – a first for Hong Kong's commercial high-rise sector.

Hysan Place is a sustainable building which seeks to make a positive environmental impact beyond its site boundary, while still contributing to the continuing commercial success of the area. It aims to address many of the environmental challenges associated with high-density and high-rise developments.

Establishing green spaces such as the Urban Farm and Sky Gardens create an entirely different building type, one that also facilitates social cohesion. Together with urban windows, these features combine to create a healthier and happier environment, one that the client hopes will influence the mindset of workers, visitors and developers in the area.

Chapter 6

Design in the digital age

Computers have been influencing how buildings are designed for many years, from carrying out vast quantities of complicated calculations to creating drawings and visualisations. This time-saving technology and ability to store a large amount of information accurately has given confidence to architects and engineers to move away from orthogonal, conventional buildings towards the possibilities of more sculptural and vertiginous forms.

Perhaps most empowering to the design team is the way "parametric modelling" is now allowing certain parameters to be safeguarded in a design, while others are allowed to vary. This leads to a final scheme that adheres to the principles most valued by its designers and which meet the architectural brief with the highest levels of structural efficiency and safety. Usable floor area for the same sized building can also increase as a result.

Building Information Modelling (BIM) is a tool that is also revolutionising the construction process by integrating the way different disciplines work. From concept stage through to detailed design, construction and facilities management, BIM provides a way of coordinating everything on a project, providing accurate visualisations and facilitating a reliable means of accommodating change. It is particularly useful for redesigning mechanical, electrical and plumbing systems to suit structural or architectural alterations. Project risk diminishes using BIM since it provides the opportunity to resolve design issues in the safe environment of a virtual model, raising the likelihood of optimum solutions to be implemented on-site.

6a
© M+ Consultancy JV

Case study 6a

M+
Hong Kong

CONTRIBUTORS: ERIC LEUNG, GAVIN TANG, ANDREW ZHANG + EDWIN-DW FUNG

Client:	West Kowloon Cultural District Authority
Design team:	Joint Venture: Herzog de Meuron (Design Architect), TFP Farrells (Executive Architect) and Arup
Arup's role:	Civil, structural, geotechnical, mechanical, electrical, public health and fire engineer, and lighting design, façade, acoustics, security and transport planning consultant
Year of completion:	2019 (expected)
Height:	100m
Number of storeys:	18
Gross floor area:	60,000m^2
Primary use:	Museum
Main contractor:	Hsin Chong Construction Group Ltd

A 40ha area of reclaimed land in Hong Kong is undergoing major transformation to create the West Kowloon Cultural District. The cornerstone for this ambitious development is a visual arts complex known as Museum Plus (M+) that will house modern art, in its broadest forms, from Hong Kong and further afield. In keeping with the new cultural quarter's contemporary and future-gazing outlook, M+ is a modern and thoughtful building whose design and construction is itself an art form.

M+ uses Building Information Modelling (BIM) as a way of calculating, visualising, and coordinating its design and construction to enable these processes to be as efficient and informative as possible (Figure 6a1).

BIM is a relatively new technique in Hong Kong and M+ is one of the few building projects to implement it from the very beginning of its design, through construction and then onto facilities management of the completed building.

M+ structure

M+ is made up of expansive gallery space measuring 110m by 130m in plan, which is bisected by an imposing, but slim, 16-storey office and restaurant block

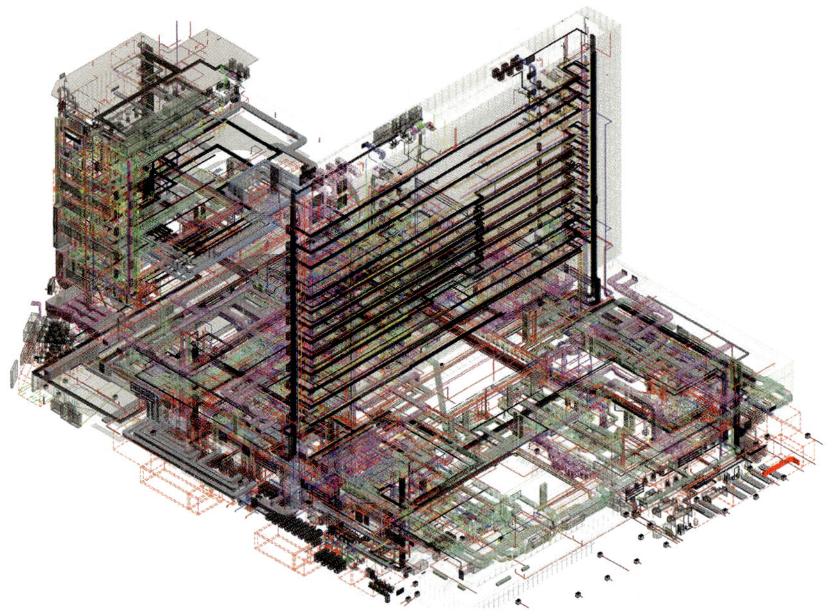

6a1
BIM: integration of architectural, structural and building services design in federated 3D models of M+
© M+ Consultancy JV

measuring 110m by 10m in plan. The podium of the building comprises a ground floor plaza and exhibition space, first floor MEP level and second floor main gallery. A basement connects this building to a separate eight-storey art storage facility (Figure 6a2).

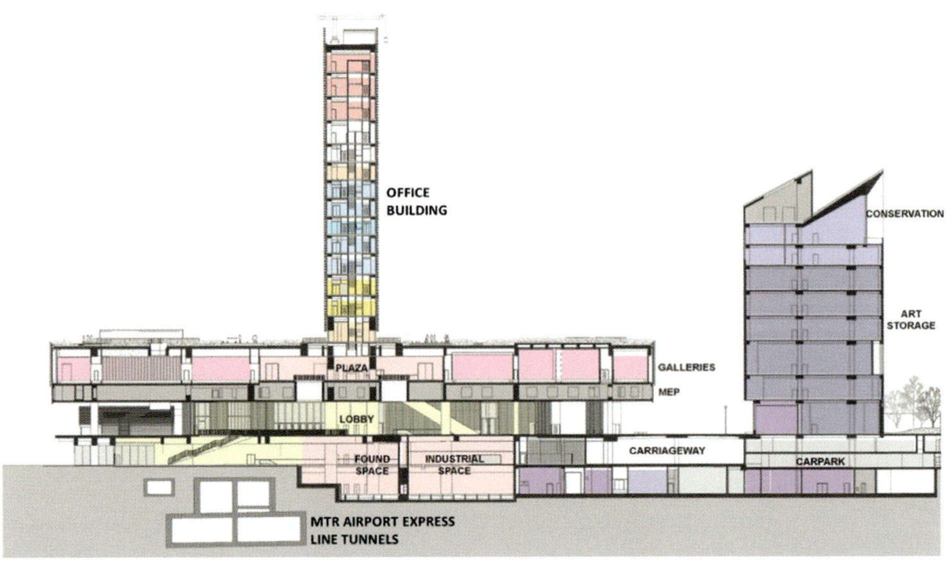

6a2
Cross-section through M+ showing types of occupancy and Airport Express Link tunnels that run directly underneath the building
© Adapted from M+ Consultancy JV

Finding the "found space"

One of the areas that could be scrutinised in more detail using BIM was the "found space", which sits above and adjacent to the MTR Airport Express Line (AEL) tunnels. The area is so-called because it was not part of M+'s original plans but was "found" by excavating into the ground so that the basement could extend right up to the MTR tunnel's boundary. The architect decided to express the presence of the tunnel infrastructure by incorporating a stepped profile in the basement, instead of simply spanning over it (Figure 6a3).

6a3
Rendered M+ gallery level model showing the stepped basement levels following the shape of the AEL tunnels
© Herzog & de Meuron

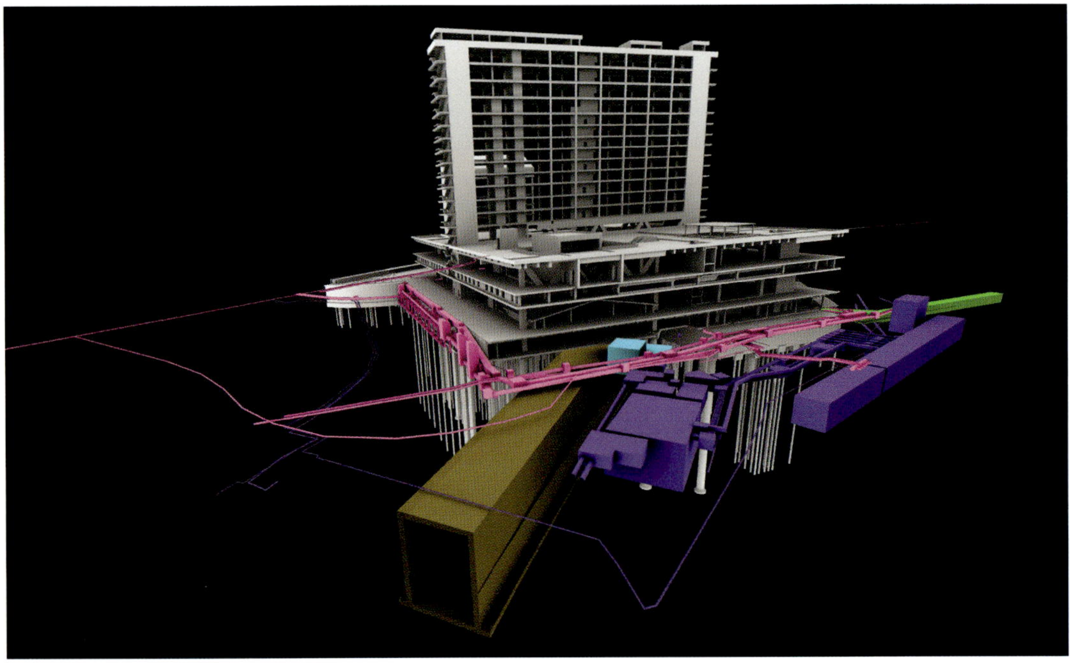

6a4
BIM model shows close proximity of buried infrastructure and basement wall and floor
© M+ Consultancy JV

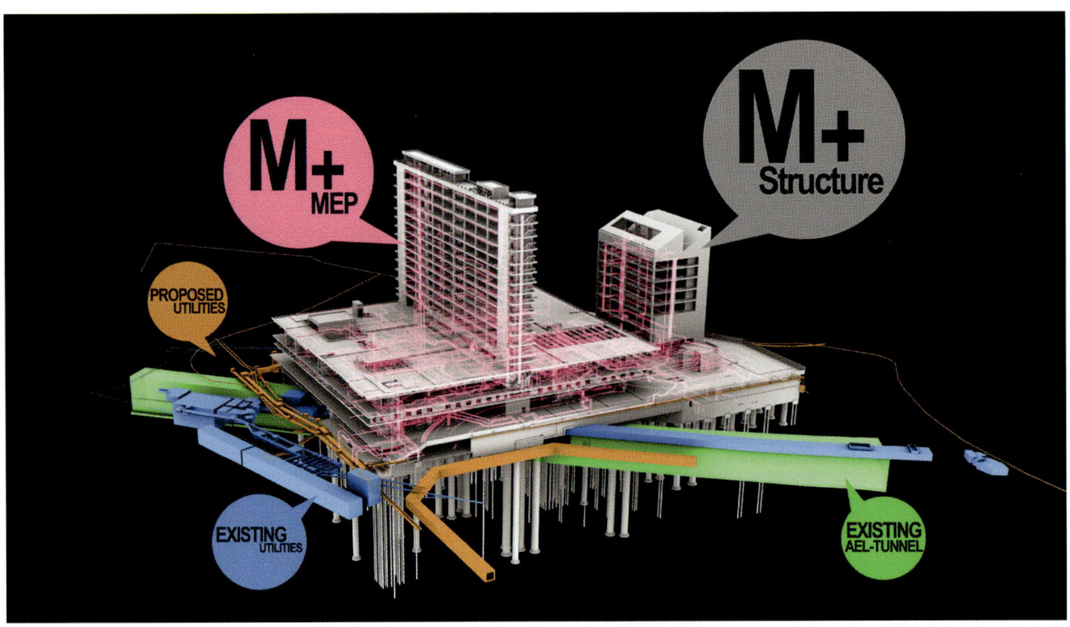

6a5
BIM allows comprehensive visualisation of all disciplines
© M+ Consultancy JV

By carefully surveying and modelling the AEL tunnels and associated services, which snake under the site, engineers could confidently define a path that allowed the loads of the building to safely by-pass this sensitive infrastructure via piled foundations placed either side of the tunnels (Figure 6a4).

Coordinating services and structure

BIM proved to be invaluable in coordinating services and structure in the M+ project with its effect of highlighting issues between disciplines earlier on in the design stage. As coordinating space requirements across disciplines is an integral part of the BIM process, this allows solutions to be worked through collaboratively and with a much greater chance of implementing the best solutions.

Located at different levels within the podium, concrete-encased steel trusses up to 60m long and 18m deep act as transfer structures to allow the building to span over underground infrastructure. They had to be carefully positioned and detailed to accommodate services passing through the building, particularly on the MEP level, which serves the gallery floors above and below. Where possible, services had to be designed to thread carefully through these trusses and, where space requirements could not be resolved, top and bottom chord reinforcement had to be detailed to accommodate openings.

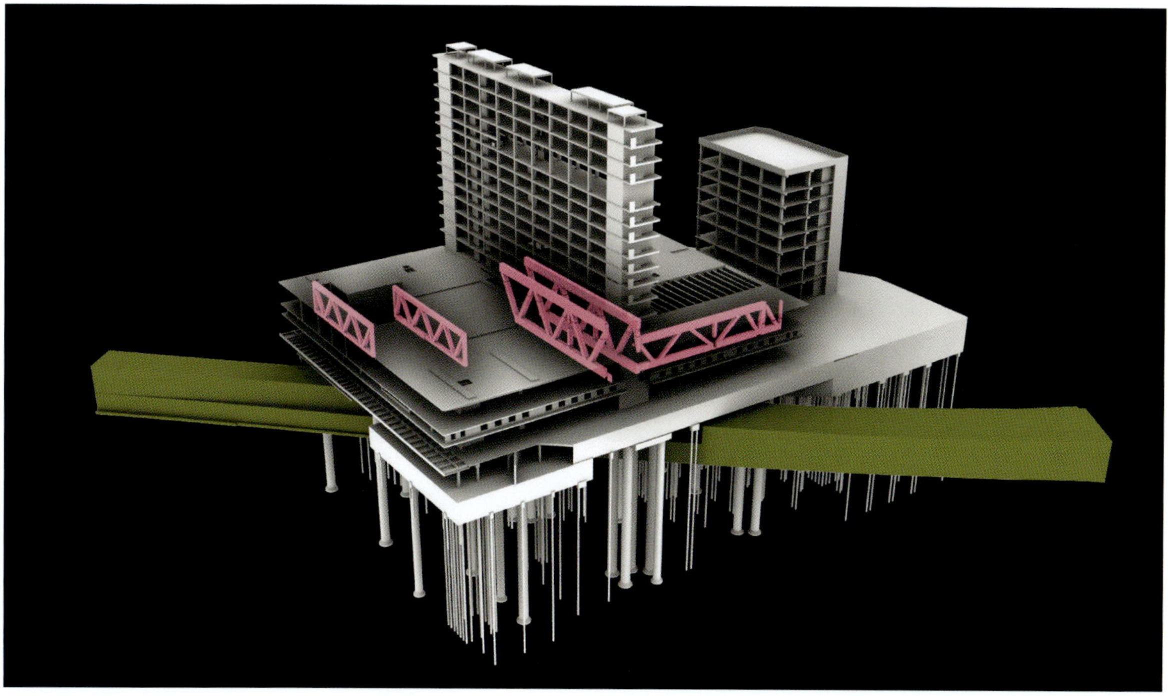

6a6
The transfer structures (pink) ensure the building does not exert excessive loads on the AEL tunnels (olive)
© M+ Consultancy JV

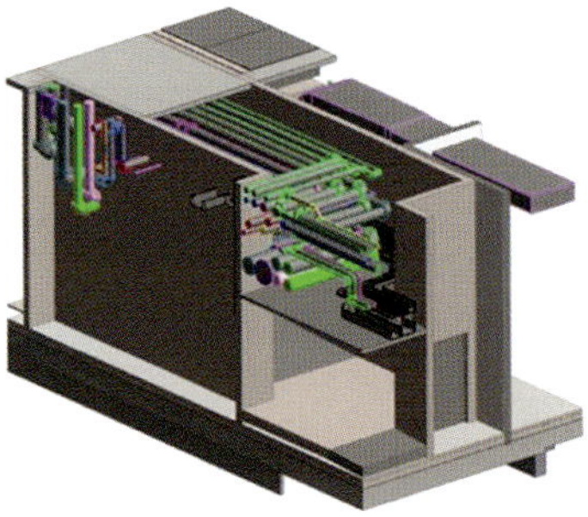

6a7
Typical BIM visualisation showing building services integrated with structure in a plant room
© Arup

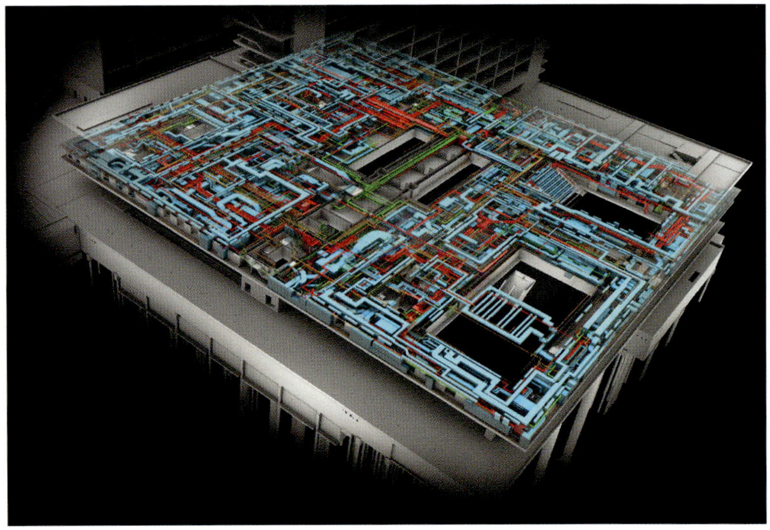

6a8
BIM model showing the MEP level with equipment in situ
© M+ Consultancy JV

The other area where BIM aided design development was in the three cores of the tower and five cores of the podium. These cores form the main stability system for the building, but are also the main routes for vertical services circulation. Services pass through wall openings in the core to each floor. These openings, however, undermine the core's stiffness, so a balance had to be struck to preserve the integrity of the core while still accommodating all the services needed. BIM helped visualise the services routes through the cores and speeded up decision-making, resulting in either openings being relocated or resized, or reinforcement in the concrete core being detailed to suit.

For the whole project, 50,000 clashes were initially detected, but this quickly reduced to 5,000 after a month. The clashes, automatically identified using BIM, consist of some that are acceptable under human judgement (such as between a partition wall and beam) and others that might require minor tweaking or fundamental changes. Solutions to resolve the latter would be decided collaboratively, led by designated architecture, structural engineering and building services BIM Coordinators, with the assistance of the 3D model in developing rerouting options. Their respective teams would later implement the changes.

Quicker MEP calculations

Information about each services conduit on the project, including its content, length, size, flow rate and material, was contained within the BIM model and could be extracted to calculate physical properties such as the loss in pressure caused by friction. Through an iterative process, these values could then be used to determine the required pump pressure for each pipe and also the most suitable pipe size. Air ducts were designed similarly, where data was automatically fed into specialist MEP software to calculate friction in the duct and airflow velocity to enable duct sizing. Different duct sizes could then be optimised by limiting the allowable friction or velocity.

Another advantage of using the BIM for MEP calculation was that changes in structure or architecture could be relatively easily accommodated since MEP calculations and routing could be automatically regenerated to suit.

A high-quality finished product, as expected

Since the project uses exposed concrete surfaces (generally without false ceilings or partitions) in the podium, there was even greater emphasis on making sure the structure and services came together neatly without any further adjustment on-site. The architect, structural engineer and building services engineer worked together, taking full advantage of being able to manipulate the size and location of structural elements and pipes in the virtual building to ensure there would be no conflict between functionality and aesthetic vision in the reality. So when construction began, the accurate rendering and sizing of elements using BIM meant that there were no major surprises for the project team or client.

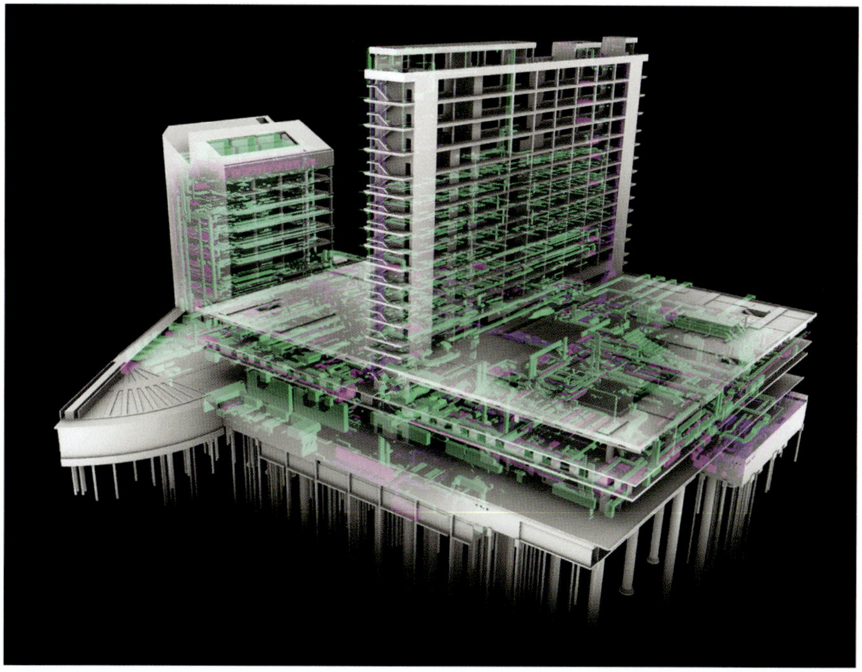

6a9
Integrated services and structural BIM model
© M+ Consultancy JV

Much more input from designers is required when BIM is used at design stage compared to a traditional approach, but there are immense gains in terms of optimisation opportunity and cost and programme certainty. And upon completion, there is a comprehensive and accurate model which can inform maintenance, refurbishment, extension and even demolition for many years to come.

The BIM process at M+

The architect typically builds a 3D model, which passes in turn to the building services engineer, the structural engineer and then the civil engineer. When the design has been developed sufficiently in accordance with the defined project BIM Execution Plan, a combined multidisciplinary 3D model is created.

The model is updated following regular coordination meetings attended by designated BIM Coordinators from each discipline, which can be as frequently as every week during the peak design period.

On completion of M+'s design, the model was passed to the main contractor to add the exact specification and supplier for every component in the building. The contractor could use the model to create virtual construction sequences and for works coordination. On completion of the building, the model will be linked to more information and used for facilities management.

6b
© Kohn Pederson Fox Associates PC

Case study 6b

China Zun (Z15)
Beijing, China

CONTRIBUTORS: PENG LIU, YU CHENG + DOROTHEE CITERNE

Client:	CITIC HEYE Investment Company
Concept architect:	Terry Farrell and Partners
Design architect:	Kohn Pedersen Fox Associates
Associate architect:	Beijing Institute of Architectural Design; CITIC General Institute of Architectural Design & Research Company
Arup's role:	Structural, geotechnical and fire engineer and security consultant
Year of completion:	2018
Height:	>500m
Number of storeys:	108
Gross floor area:	437,000m^2
Primary use:	Office
Main contractor:	China State Construction Engineering Corporation

The elegant, curved elevations of the China Zun tower are a result of delicately balancing architectural form, structural efficiency, safe construction and commercial needs. While many buildings might profess to be designed with these aims, the Zun can demonstrate it actually has through its use of parametric modelling.

The method of design involves using computer software to systematically work through every iteration possible for a building's geometry, which adheres to rules set by the architect's sense of aesthetic form and structural engineer's sense of logic to achieve that form. These rules relate to aspects such as maximising floor area, optimising structural efficiency or facilitating simpler construction. By then analysing these possible solutions and refining them further, a much more efficient building can evolve. Rational, high-level decisions are still made by human minds, but the time-consuming task of number crunching is carried out digitally.

Why parametric modelling?

Parametric modelling was used to design the Zun for three reasons. The first was because the architect needed to explore many different options to optimise its overall form to meet planning requirements and because each option needed to be quickly analysed by the structural engineer. Using traditional analysis methods, each option would take two engineers at least seven days to complete. Using parametric modelling, it took just two hours.

Secondly, due to the Zun being much taller than any other building located in a high seismic area, there were no examples to offer guidance for its design. Parametric modelling allowed different options to be compared, providing greater confidence in going ahead with the final scheme.

Lastly, the shape of the tower and its curved façades required complex and lengthy structural analysis. To aid stability, most tall buildings tend to have a constant cross-section or have one that diminishes with height. But the Zun, taking its inspiration from a Chinese water vessel with a narrow waist and enlarging upper portion, has inherent weaknesses, exacerbating the challenge of building such a tall tower in an earthquake zone. Connecting parametric models to structural analysis software (using an Arup-designed program) allowed the effect of changing the architectural form to be analysed accurately.

Perfecting the geometry of the building's curved façades to project elegance and engender a robust structural design was the result of processing hundreds of versions of the scheme. Four are shown in Figure 6b1.

The main structure

The tower is square in plan with chamfered corners and has a central reinforced concrete core and perimeter steel braced frame. This perimeter steelwork is made up of gravity columns; belt trusses, which visually divide the height of the building into zones; mega columns, which run the full height of the building; and mega cross-bracing (Figure 6b2).

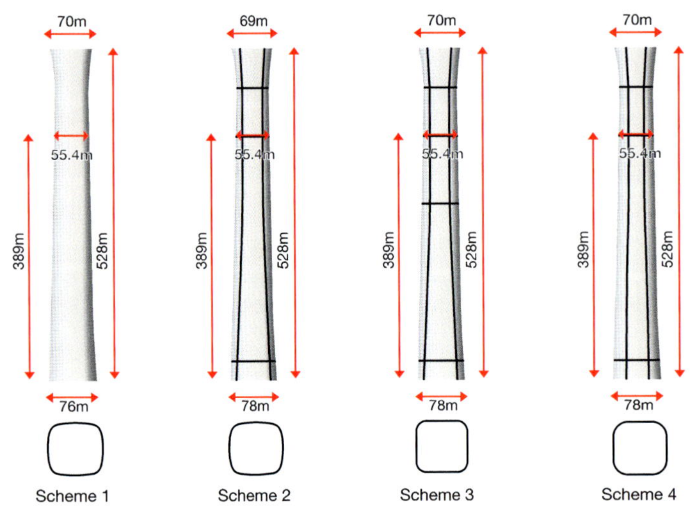

6b1

Four of the different architectural options investigated using parametric modelling. Scheme 1: initial model; Scheme 2: all sides of the tower plan are arcs; Scheme 3: an extra shape control level is added and all sides of the tower plan are straight lines; Scheme 4: all sides of the tower plan are arcs and the fillet radius is increased © Arup

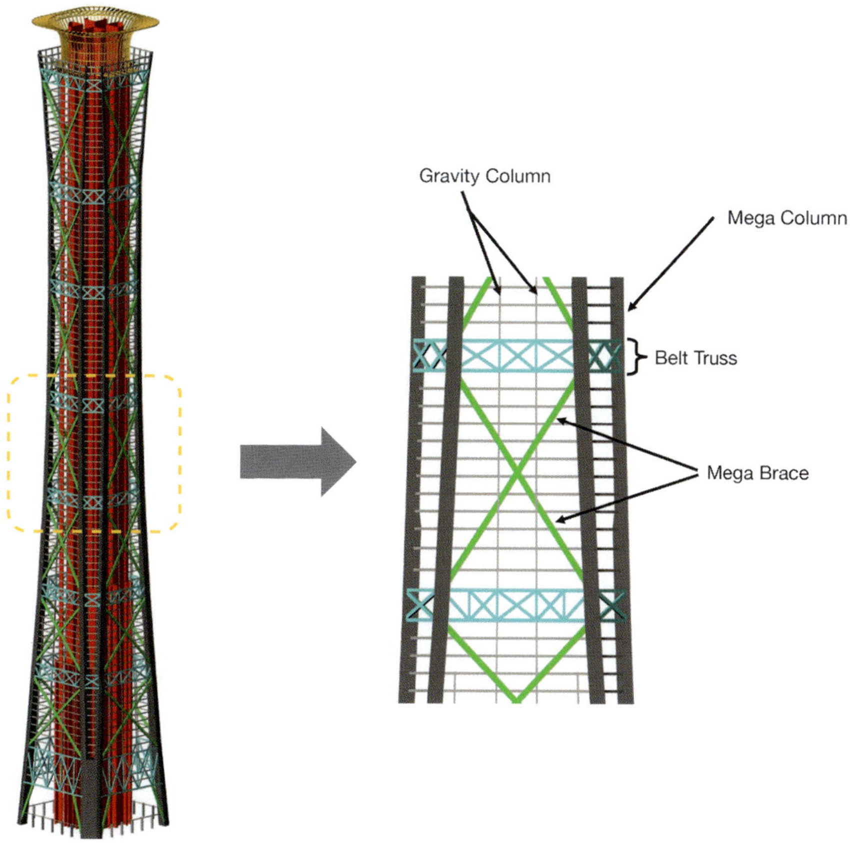

6b2
The Zun structural system
© Arup

The mainstays of this 108-storey building are its mega columns, which sit in each corner of the tower. Up to level seven, there is one column per corner, but above this, a pair of columns is required. Although they appear to follow the contours of the façade, they are in fact straight lengths that change angle at various intervals to create the illusion of a curve.

Perfecting the aesthetic form

Optimising the form of the tower involved varying the following four aspects (Figure 6b3):

- width of the square floorplate (H and B)
- levels at which the radius of the façade arc changed to achieve the required curved profile (control plan level, CPL, 1 to 5)
- radius of the chamfered corner of each floorplate (R)
- radius of the façade's arc (r)

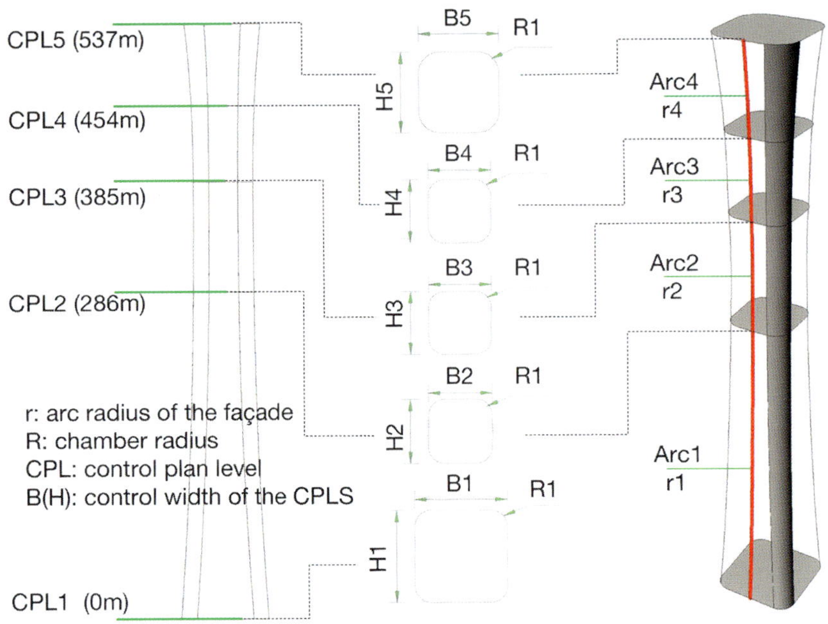

6b3
The geometric control parameters of the tower
© Arup

Optimising the mega column location and profile

Since the mega columns are so important to the structural stability of the tower and all perimeter steelwork is positioned relative to them, determining their geometry unlocked many of the complexities of the Zun. Parametric modelling was used to set rules, or constraints, which influenced the mega column's design. They included investigating options to minimise the distance between the mega column and the curtain wall. This would maximise usable floor area as this gap would be occupied by perimeter steelwork. While adhering to this rule, another was created which allowed sufficient room for operatives to install the curtain wall. Keeping the distance between mega column and curtain wall as small as possible also had the added benefit of maximising the building's lateral stiffness (by maximising the lever arm between mega columns). This would be beneficial for satisfying seismic and wind code requirements.

However, the desire to minimise the distance between mega column and curtain wall would also lead to more connections required in the column to allow it to follow the curve of the façade more closely. This would be at odds with reducing the number of connections in the mega columns and facilitating simpler construction. The parametric model was able to work through these constraints to produce a building geometry that satisfied all these needs as far as possible.

Analysing design constraints

Arup's parametric model was investigated further using structural analysis software to understand the effect each different architectural option had on the

6b4
Constraints on the position of mega columns
© Arup

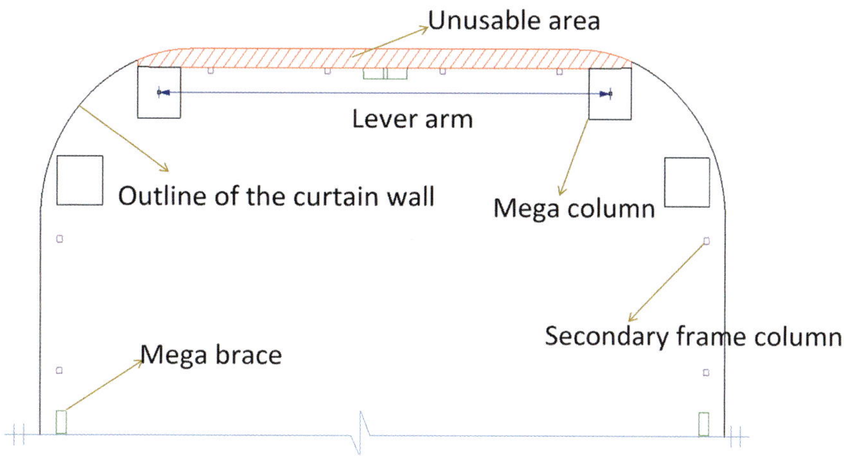

6b5
Relationship between the centerline angles α, lever arm (see Figure 6b4) and unusable area
© Arup

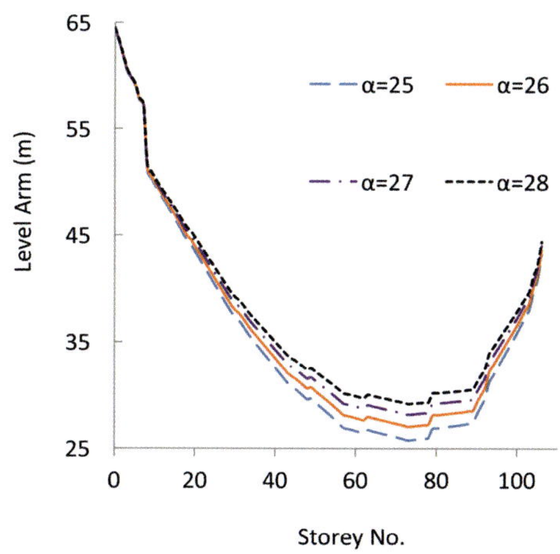

α	Unusable area
25°	12,975.3 m^2
26°	14,514.1 m^2
27°	16,178.5 m^2
28°	17,943.5 m^2

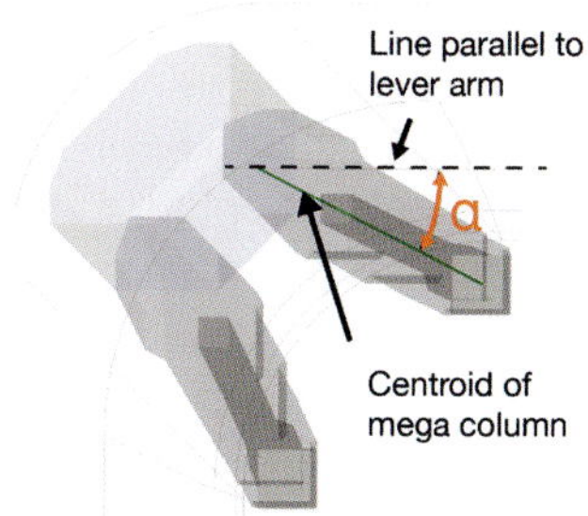

structure. This required results from the parametric model to feed into structural analysis software with the addition of material, section size and loading information. The analysis assumed that the mega columns would be connected to the belt trusses to some degree along their length. Each connection would also accommodate a change in vertical angle of the mega column, β (Figure 6b6). The analysis also looked at whether each mega column should be tied to both the upper and lower chords of each belt truss or just one of the chords.

Each scenario would affect the proportion of lateral load transferred into the belt truss members, and greatly influence the sizes they would need to be. The analysis revealed that the most favourable loading arrangement would be achieved by connecting the mega column to both the upper and lower belt truss chords above and below the seventh zone only. Elsewhere, they would only need to be connected to the lower chord. This was the best scenario to keep the distance between façade and mega column as small as possible, limiting the number of

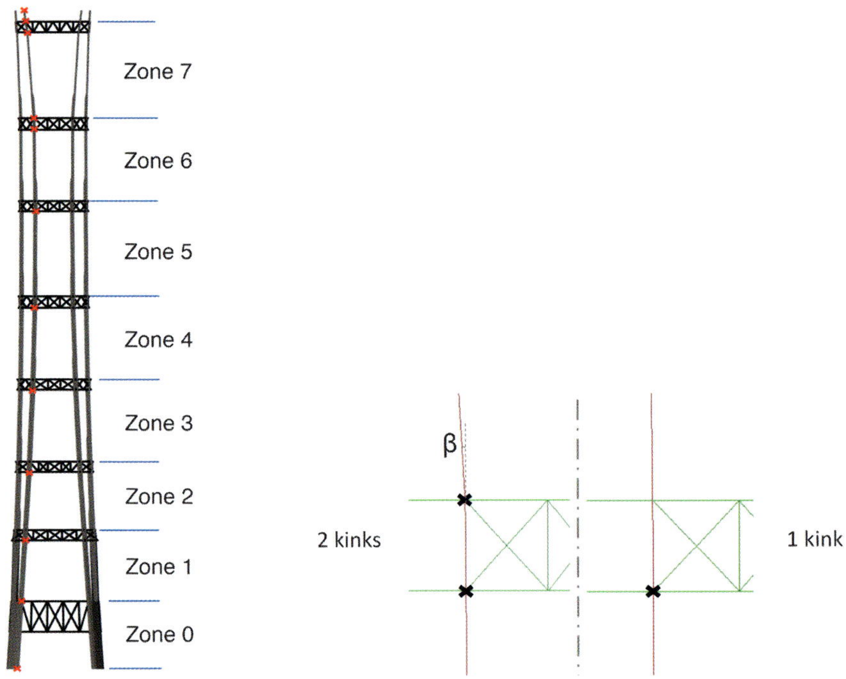

6b6
Twelve connection points between the belt truss and mega column (red crosses) and the "kink" angle, β, limited to 6°
© Arup

connections to as few as possible (12) as well as rationalising the forces transferred into the belt truss. The analysis limited the "kink" angle, β, to 6°.

Other design constraints included keeping the overall mega column cross-section size as constant as possible over the height of the building to avoid the tendency for its perimeter welded plates to twist.

Complying with building codes and achieving cost savings

This structural analysis model was scrutinised further using additional software to satisfy wind, seismic and additional structural design codes. Costs were also evaluated by assessing the tonnage for each option. Critical details were extracted from the model for finite element analysis. This was one of the first times Arup had executed a smart design framework for a project, taking in parametric modelling, structural analysis and cost analysis, which then resulted in a final Building Information Model (BIM) which could be shared by other parties for the detailed design.

Optimising the geometry of the Zun using parametric modelling resulted in gaining an extra 7,800m^2 of floor area. Comparison of a partially braced tower using heavier sections and a fully braced tower with lighter sections also revealed that the smaller sections in the fully braced design would lead to 50kg/m^2 lower steel tonnage.

The use of computer technology has minimised design and construction risks, optimised material costs and maximised the useful floor area to create a stunning building. The Zun is as close to perfection as is humanly and technologically possible.

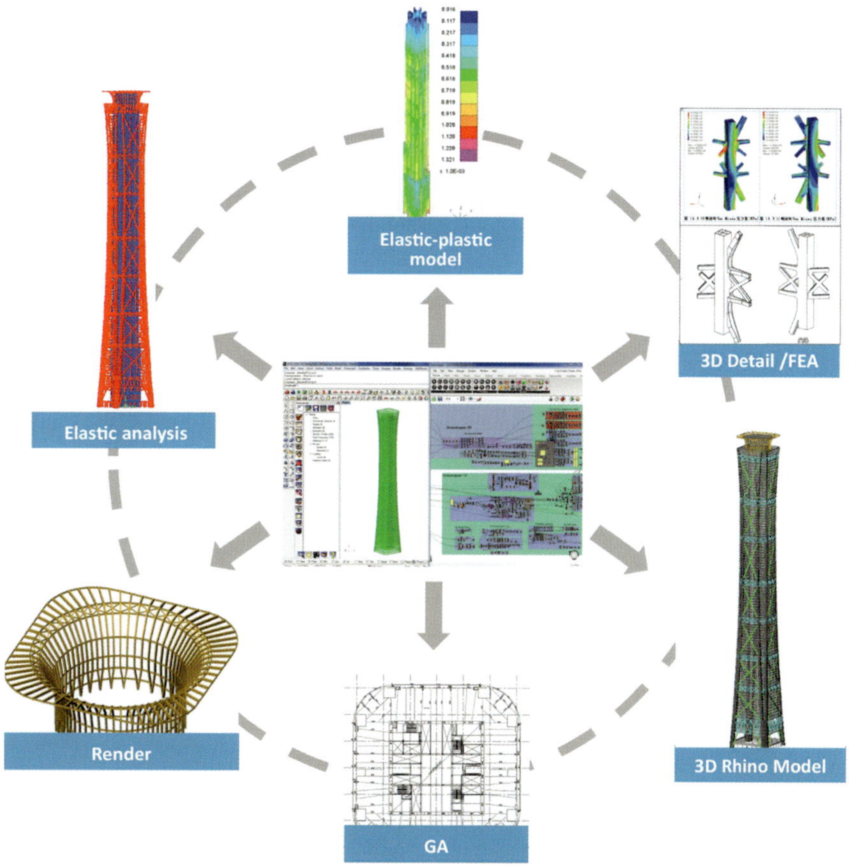

6b7
The interoperability enabled by the smart design framework in this project
© Arup

Parametric modelling for complex structural and façade systems

Just 250m south of Z15 is Z6, a 405m tall tower also designed by Arup using parametric modelling. Z6's structural diagrid system is integral to its façade, so refining its structure had a profound effect on how the building would look. The diagrid's final, optimised arrangement was derived by carefully balancing:

- Overall structural behaviour and requirements at key connections
- Variation in different cladding panels (fewer variations would lead to reduced cost)
- Overall aesthetics of the façade.

Apart from using parametric tools to generate different options for the main structural members of the diagrid, Arup also developed a second set of tools to provide a rapid and visual appraisal of each diagrid scheme. This involved measuring the length, angles and number of intersections of all members. Summarising this visually enabled rapid comparison of different schemes and helped highlight areas that would need further refinement.

Figure 6b8
A pattern of mega and mini braces curving in different directions form Z6's structural diagrid and its distinctive doubly curved shape
© Foster & Partners

Chapter 7

Total Design

Total Design is a philosophy developed by Arup's founding father Ove Arup to promote the benefits of integrated working. The underlying theory is that a single discipline working in isolation cannot deliver the best possible solution and that collaboration is essential to producing better, more holistic designs. This is particularly important in tall buildings, where many skills are needed to develop the vast networks of structural, mechanical, electrical and plumbing systems that allow these vertical cities to function efficiently.

Total Design brings into focus the synergies that already exist between many disciplines to encourage optimum results for all. To unlock its full potential, it needs to be implemented from the start of a project and continue to completion. Multiple like-minded firms can achieve this, but it is most effective when a single firm can offer the many technical skills needed to deliver a project.

As Arup's services have diversified, undertaking many different roles on the same project has become feasible. Office expansion across the globe, coupled with advances in IT, are now allowing specialist resources from around the world to work seamlessly together.

For the most complex tall building projects, collaboration is often the most powerful design tool, leading to simple, logical and safe solutions – and making sure that the most out-of-this-world ambitions can be fulfilled.

7a
© Kenny Ip

Case study 7a

Canton Tower
Guangzhou, China

CONTRIBUTORS: TONY CHOI, GEORGE ZHAO + JAMES CHEUNG

Client:	Guangzhou New TV Tower Construction Co. Ltd
Design architect:	Information Based Architecture
Architect of record:	Guangzhou Design Institute
Arup's role:	Lead consultant for architecture, master planning, structural, mechanical and electrical engineering and cost control of the design phase
Year of completion:	2010
Height:	600m
Number of storeys:	37
Gross floor area:	114,054m^2
Use:	Telecommunications and visitor attraction
Main contractor:	Guangzhou Municipal Construction Group JV; Shanghai Construction Group

Spiralling 600m into the sky above Guangzhou is the Canton Tower, a symbol of the city's cultural and economic success. Elegant, functional, yet playful, this landmark is made up of steel tube columns that taper and twist around an eccentric elliptical reinforced concrete lift and stairs core.

A lattice tube structure made up of diagonal steel members brace these columns while another layer of structural and architectural dynamism is created by elliptical steel rings sloping at 15.5° to the horizontal that stretch across the full height of the tower. The degree of twist of the columns is defined by a rooftop ellipse measuring 40.5m by 54m, which is orientated at an angle of 45° to the tower's larger 60m by 80m elliptical base.

Concealed behind this exoskeleton are the functional spaces: 37 occupied floors parcelled into five blocks and separated by zones of open lattice structure. An observatory is located on the uppermost floor, while restaurants, television studios, visitor amenities and a cinema occupy others.

The tallest freestanding structure in the world when completed in 2010, this tower was designed to appear complex, but, in fact, be simple to build. This was achieved through its "Total Design" solution, where all structural engineering, building services, wind, seismic, fire and lighting design was developed holistically by a single firm, Arup, in close collaboration with the architect from day one.

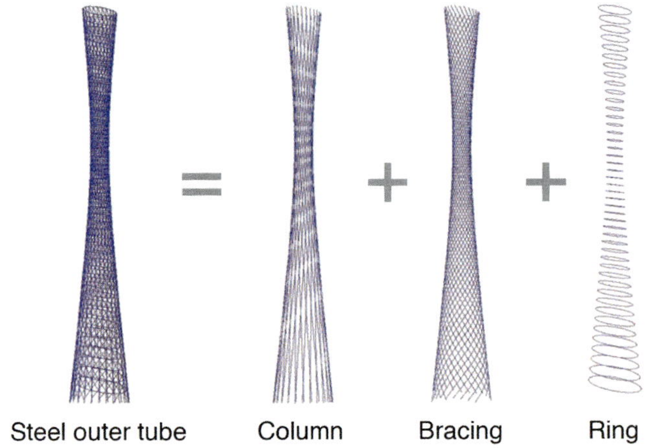

7a1
Exoskeleton structure comprising columns, rings and diagonal bracing
© Arup

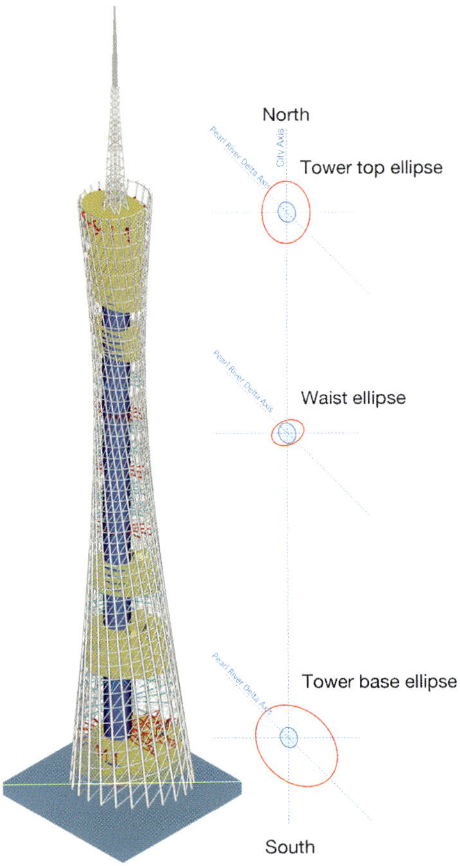

7a2
Eccentrically positioned elliptical services core, superimposed onto the cross-sections of the tower at its base, waist and roof
© Arup

A balance of form and function

The architect set the starting point for the two ellipses that geometrically define the tower's base and roof, but, beyond this, the iterative process of determining the number and degree of twist of columns, and number and angle of sloping rings to achieve the desired form, was supported by Arup using parametric modelling.

However, nothing in this tower is designed in isolation, so in conjunction with developing the tower's geometric form, other design- and construction-related criteria needed to be considered to ensure the balance between form and function was maintained. Arup eventually reduced the number of columns from 30 to 24, which also reduced the number of connections to make the structure simpler to build. The columns were also designed to be concrete-filled to improve their fire performance, add weight to the lightweight tower (to help "anchor" it down) and to provide increased stiffness.

Further optimisation to take into account other stress factors influenced steel plate thickness, which varies from 30mm to 50mm, and the taper of the columns, which reduces from 2m in diameter at the bottom to 1.1m at the roof.

Detailed design of other components including 46 rings and diagonal bracing were developed through parametric modelling. Their specific sizes, thicknesses and spacing were fixed following seismic and wind analysis, which identified areas in the tower that needed extra strengthening.

The waist: fit and slim

The slim waist caused particular structural concern and is fundamental to the tower's elegance. It was only near the end of the design process that the tower's waist dimensions could be finalised to just over 22m wide at its narrowest. This was the result of much cross-discipline collaboration to strike a balance between functional, architectural and budgetary needs with due consideration to the complex loading combinations, requiring critical wind, seismic and thermal loadings to be modelled.

The waist dimension is also highly dependent on the final core size, an ellipse measuring 15m by 18m.

The core has to accommodate six lifts, minimum areas for services risers, lobbies, staircases and the required number of exits. Waiting areas and vertical circulation routes were based on normal operation mode as well as during an emergency, and so had to correspond with the tower's fire evacuation strategy. To keep the core compact and slim, services ducts had to be sized efficiently and features such as double deck scenic lifts incorporated instead of single ones to transport twice the number of people in the same plan area.

To provide adequate rigidity to this narrowing, the rings are spaced just 8m apart nearest the waist (they are 12m apart nearer the top and bottom) to make the angle of diagonal bracing more structurally effective.

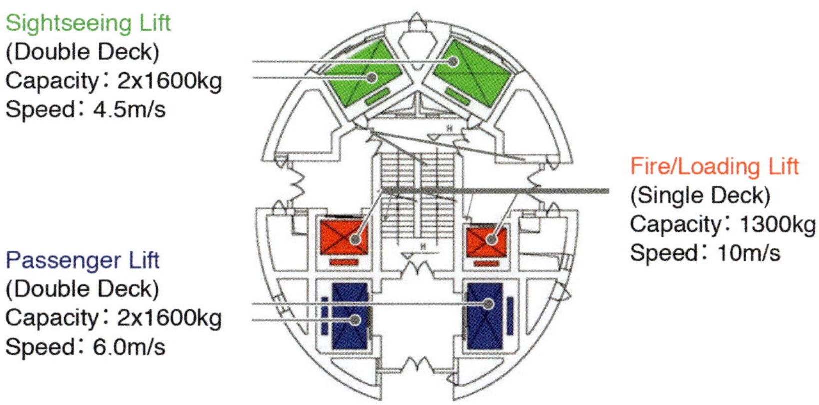

7a3
Compact core showing lifts, stairs and lobbies
© Arup

The wind that blows through it

Since the tower is a combination of an open structure where the wind can pass through (much like an electricity pylon) and a closed one around the occupied floors, its wind forces are difficult to assess using code-prescribed methods. The open portion, for example, experiences complex wind loading patterns, as these forces effect many surfaces – the facing structure, the core behind it and also the inside face of the structure on the opposite side.

Wind tunnel tests were carried out to determine the wind loading that would be experienced by the structure, as well as its acceleration under these loads. The topography model was built to a 1:2,000 scale, but due to the complex nature of the tower, a larger 1:150 scale model was also built to assess the wind effects on each of the nine sections of the model.

With the wind loadings derived from the tests, engineers had to design the structure to the levels of comfort and safety required under normal and more extreme scenarios and for degrees of movement considered tolerable to the public and staff. Combined with the fact that the tower would be closed to the public during strong winds, but might require TV transmission staff to remain, the benchmarks for allowable wind-induced movement across the height of the tower could then be developed and the structure designed accordingly.

Standing firm

As a unique hybrid structure – part mast and part building – standard seismic design methods could not be applied to the Canton Tower. The criteria for limiting

7a4
Wind tunnel testing of the tower at 1:2,000 scale (below) and 1:150 scale (right)
© Arup

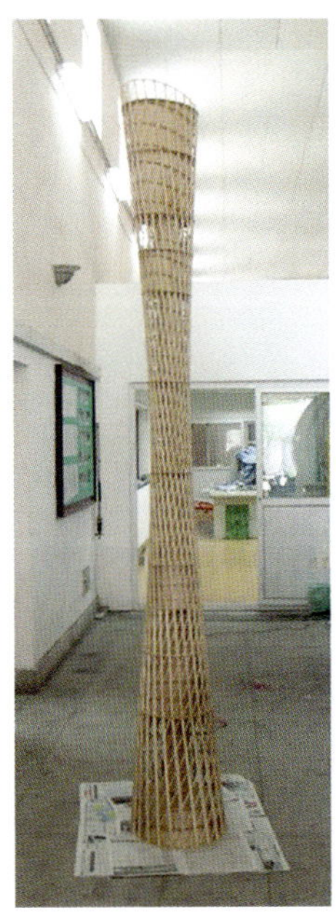

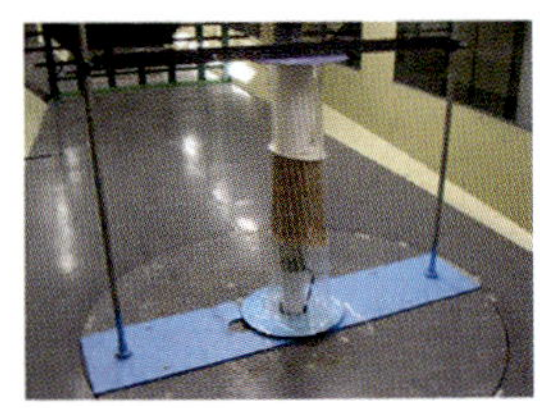

inter-storey drift would, therefore, be based on the performance requirements of the structure, building services, fire control and TV transmission, as well as occupant comfort.

For low and medium intensity earthquakes, the performance-based criteria stated that tower occupants should be unaffected and the structure undamaged.

7a5
Shake table test of a 1:50 tower model
© Arup

For rare earthquakes (1 in 2,475 years return period) the structure was required to remain intact, or "no collapse of the tower body and no breaking of the antenna".

Computer models were used to understand how the structure would be affected by these earthquakes and, in the most severe scenarios, some members were strengthened or redesigned accordingly. The overall results were verified by shake table test.

Solar and wind speed effects

The tower's structure is exposed to solar heating and susceptible to uneven expansion due to temperature effects. A great deal of work was carried out by Arup's building physics engineers to understand how and when the angle of the sun's rays would affect every structural member. Analysis revealed that temperatures of 70°C could be experienced on one side of the tower in direct sunlight, compared to 33°C on the opposite side in the shade, causing the tower to bend. The effect of solar gain imposed stresses in the structure that had to be resisted through detailed design.

The phenomenon of greater wind speeds higher up the tower also had to be taken into account in the building services design. This is because materials exposed to higher wind speeds radiate heat faster, leading to lower indoor temperature. To understand the anticipated temperature variation, building services engineers divided the 450m tall tower (excluding the mast) into seven zones and calculated the heat transfer coefficients of different building materials for each one. The results showed that when the temperature on the ground was 33.5°C, it would only reach 30.6°C at the top of the tower. With the recommended indoor temperature being 26°C, significant energy savings could be made on the uppermost floors where the air would need to be cooled down to a much lesser extent than on lower floors.

The lower indoor temperature further up the tower also inspired a building services solution which adopted two different types of air-conditioning systems in the tower: traditional water-cooled air-conditioning for the bottom 150m, with a low-energy air-cooled air-conditioning system for the top 120m.

Raising it up

Construction methods were considered in parallel with structural design, particularly in devising the relationship between the columns, diagonal bracing and rings. The rings were located on the inside face of the columns and connected via a stiffening joint. This had practical benefits in creating space to comfortably and accurately make the connection between ring and column, but also had aesthetic advantages of diminishing the visual impact of the rings from the outside and accentuating them when viewed from the inside. Bracing was positioned in the plane of the columns, which aided the development of prefabricated intersections.

7a6
Viewing the tower from the inside and outside: rings dominate the internal view (left) and twisting columns dominate the external view (right)
© Arup

A detailed constructability study was carried out at the early design stage to establish the particular construction specifications and thus the detailing of steelwork to suit.

With the decision made early on to build the reinforced concrete core using self-climbing formwork, two tower cranes could be mounted on the core for all lifting operations while core construction proceeded. To reduce the number of pieces to be lifted, as well as the number of complex connections to be made at height, the intersections of column and diagonal bracing were prefabricated. A single node-type was developed for these intersections that could be adapted to suit the different column and bracing angles and sizes.

This reduced the construction time and cost compared to designing each node individually and saved time on-site as only one set of installation and checking procedures was needed for all 1,104 nodes. The column segments with the prefabricated nodes could be lifted and positioned easily and prepared for ring and bracing extensions to be subsequently welded on.

From the main roof, the lower part of the 150m high mast, which is of lattice frame, was installed by a tower crane while the top 97m section was installed by the jacking method.

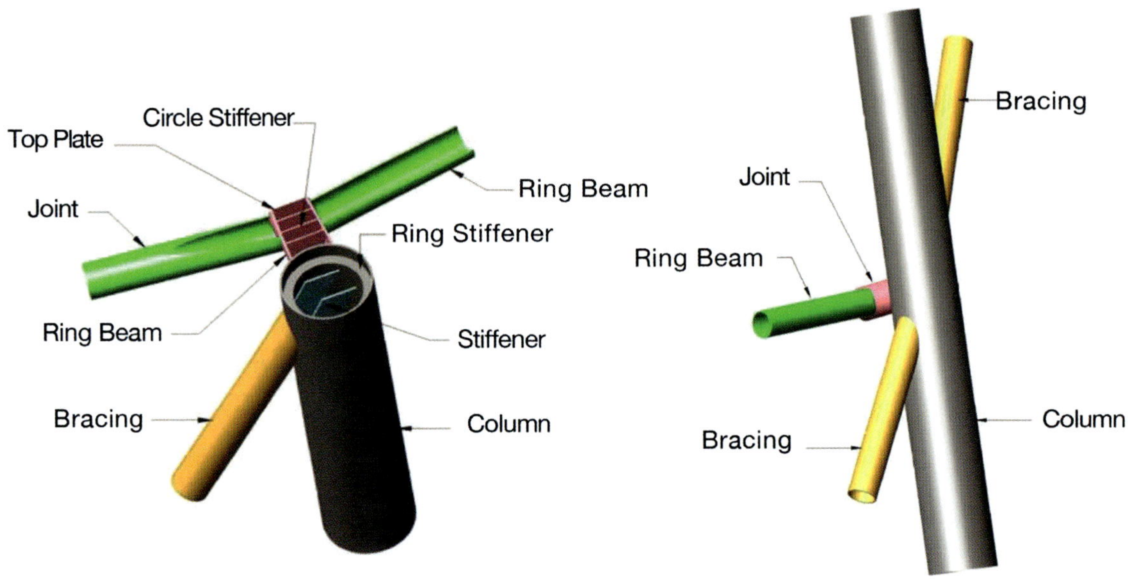

7a7
The standard node, showing how it was adapted to suit different locations
© Arup

7a8
Erection of the prefabricated column and bracing intersection
© Arup

Water tank damping

"Total Design" is perhaps best exemplified by the two water tanks that feed the tower's fire safety sprinkler and fire hydrant systems. Weighing 600t in total, these tanks, located at a height of 438m, are mounted on rollers and act as a passive damping system together with a 50t active mass damper to counter the sway and potential shaking caused by high winds. During high winds, a sensor picks up movement in the tower and activates the tanks to move in the opposite direction, reducing the shaking amplitude by 40%. The damping effect of the water tanks also contributes to improving the tower's response in an earthquake.

Locating the water tank at the top of the building also provided a direct and safe fire extinguishing solution, as water could flow under gravity and did not need to be pumped, especially important in an emergency when there may be no electricity supply. As an extra precaution, however, auxiliary water tanks are located on other floors.

Total Design that goes beyond design

Canton Tower's design programme was very fast-paced to ensure the tower would be ready to transmit the 2010 Asian Games, so research, analysis and approvals were carried out in a streamlined fashion to influence design decisions without causing delays. The "Total Design" methodology encouraged deeper collaboration across all disciplines to ensure the tower was delivered to the highest possible standard.

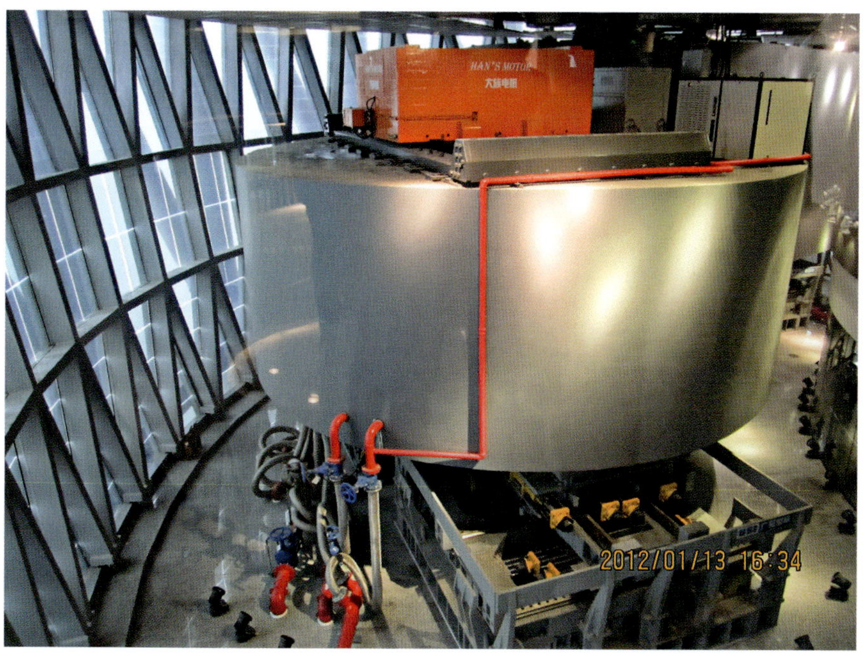

7a9
The Canton Tower's water tank damping system
© Arup

The influence of "Total Design" goes far beyond. For example, the building movement monitoring devices, which were installed at the later part of the construction stage, are still serving the operating tower today, monitoring movement effects of drift, acceleration and stress due to wind, temperature and seismic forces. The automated system with movement sensors and data collection and processing devices can visualise data and therefore enable on-site and remote data management, greatly contributing to a smart tower in operation.

7b

Case study 7b

Marina Bay Sands
Singapore

CONTRIBUTORS: VA-CHAN CHEONG + BRIAN MAK

Client:	Las Vegas Sands Corporation
Design architect:	Moshe Safdie & Associates Ltd
Executive architect:	Aedas Pte
Arup's role:	Structural, geotechnical, civil, and fire engineer, and façade, risk, security, acoustics & audiovisual, and transport planning consultant
Year of completion:	2010
Height:	206.9m
Number of storeys:	57
Gross floor area:	581,400m^2
Use:	Hotel, conference, retail, leisure
Main contractor:	SsangYong Engineering and Construction Co. Ltd (Hotel); JFE/Yong Nam JV (SkyPark)

Marina Bay Sands is an awe-inspiring business and leisure destination in Singapore, built on 15.4 hectares of land reclaimed from the sea. The site hosts a 2,560-room hotel, extensive conference, exhibition, shopping and retail areas, two theatres, casinos, a museum and twin water-bound "Crystal Pavilions" (Figure 7b1). Designing the resort was extremely challenging, due to its size, many technical complexities and because it was programmed for completion in just five years.

Arup provided many of the design skills that helped deliver the scheme and its ambitious construction programme, coming up with solutions that addressed its challenges holistically.

The Marina Bay Sands hotel is the resort's most prominent building – a spectacular complex made up of three interconnected towers with sloping walls topped by a stunning 1ha park. The "SkyPark's" viewing platform is the longest public cantilever structure in the world.

Leading civil, structural, façade, fire and wind engineering on the SkyPark and responsible for dynamic analyses and 3D modelling, Arup's multidisciplinary role ensured that the final design was well integrated – a "Total Design".

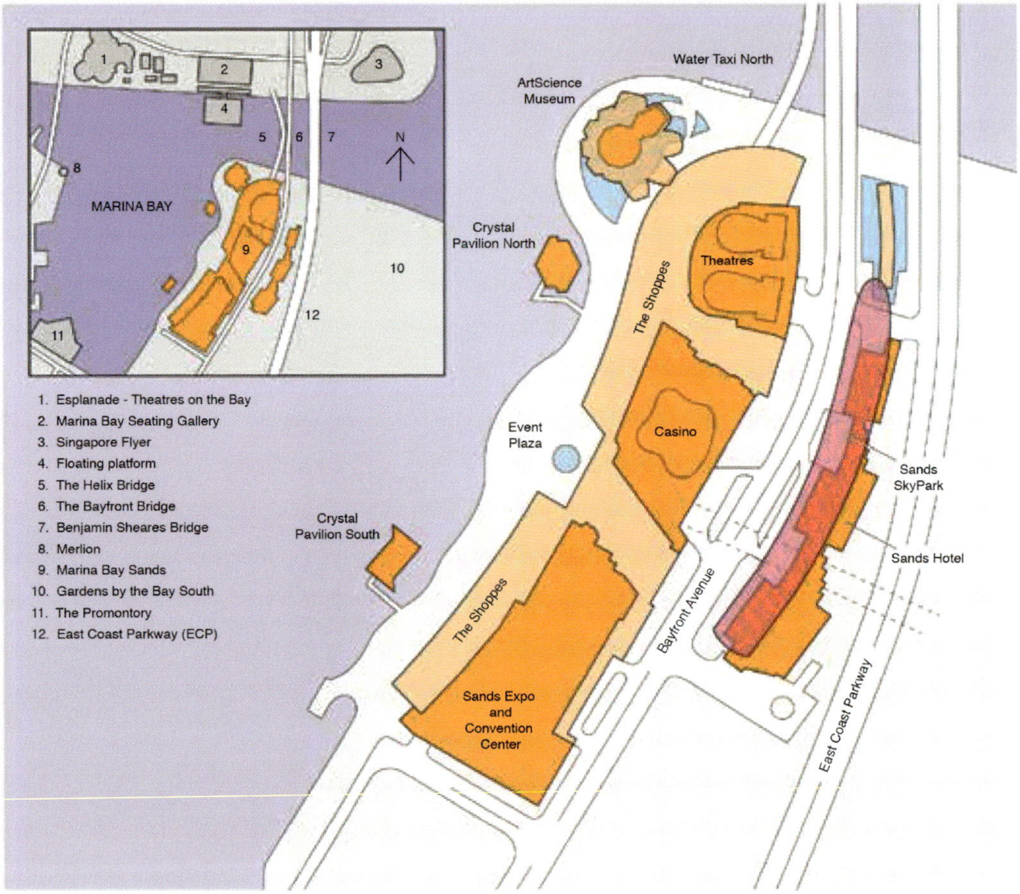

7b1
Location plan and site plan of Marina Bay Sands resort
© Nigel Whale/Arup

The park in the sky

With limited room on the ground for greenery and open spaces, the architect took the opportunity to place a 38m wide by 340m long park on the roof of the hotel. This "SkyPark" bridges the 50m gaps between the Marina Bay Sands hotel towers and oversails the edge of Tower 3, the northernmost tower, by 66.5m. Around 500 trees up to 8m tall, a 146m long infinity pool and a viewing platform make this rooftop oasis one of the most spectacular city experiences in the world.

Developing the SkyPark concept into a safe and robust structure required some clever thinking and, without "Total Design" pulling distinct disciplines in the same direction, the SkyPark may never have been feasible. With so many services led by Arup, collaboration was a very natural part of design development, yielding optimum solutions for this unique structure.

7b2
The infinity pool at the SkyPark
© Timothy Hursley

7b3
The SkyPark spanning across three towers
© Andy Gardner/Arup

The SkyPark's challenges were numerous: to design the structure to be both robust and lightweight and exert an acceptable level of loading on the towers; to accommodate movement in an architecturally acceptable way; to address the dynamic and aerodynamic cantilever issues; and to ensure noise did not transfer to the hotel. It would also need to be built safely.

Unlike the reinforced concrete framed towers, a steel frame was chosen for the SkyPark to take advantage of steel's high strength to weight ratio and to tackle the long spans. Customised steel tubular columns connect the SkyPark to the towers (Figure 7b4).

7b4
Customised steel tubular columns connecting the SkyPark to the towers
© Timothy Hursley

Accommodating movement

Although all three towers are of the same height, they are all slightly different in shape. Their sloping walls induce extra strains and movement in the structure (see box: A trio of towers) and long-term creep and settlement is not constant across the towers. To absorb this movement, as well as that induced by earthquakes, the SkyPark is designed as five discrete sections, articulated by four sets of joints with bearings (Figure 7b5). This allows each tower and the "hammerhead" portion of SkyPark to move as one. Arup explored many design options before coming up with this solution, which also offered advantages for construction.

The SkyPark structure consists of a steel frame with composite floor slab above Towers 1 and 2 and a set of three, braced 50m long longitudinal steel trusses bridging the gap between the towers (Figure 7b5). Located on top of Tower 3 is the final section of the SkyPark, which cantilevers 66.5m at a height of 200m above ground. The central lift core of each tower penetrates the SkyPark to provide additional stiffness.

Short-term movement due to wind and the dynamic response due to people moving on the cantilever created further challenges for designers.

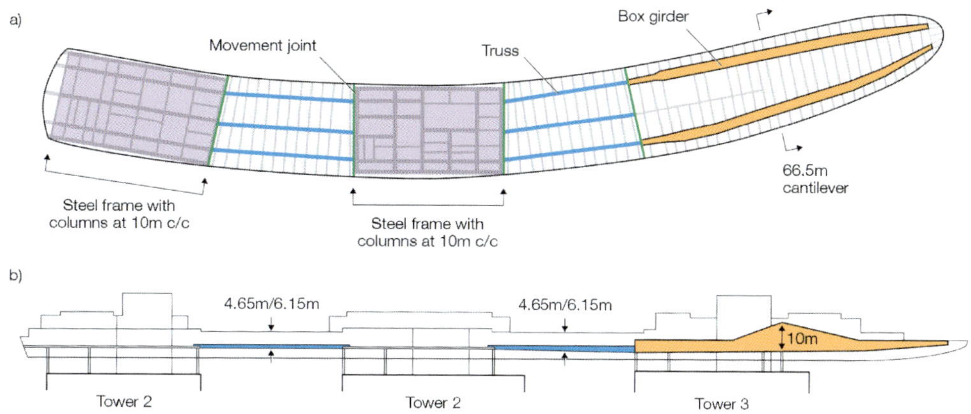

7b5
The SkyPark is split into five sections that are connected by movement joints
© Nigel Whale/Arup

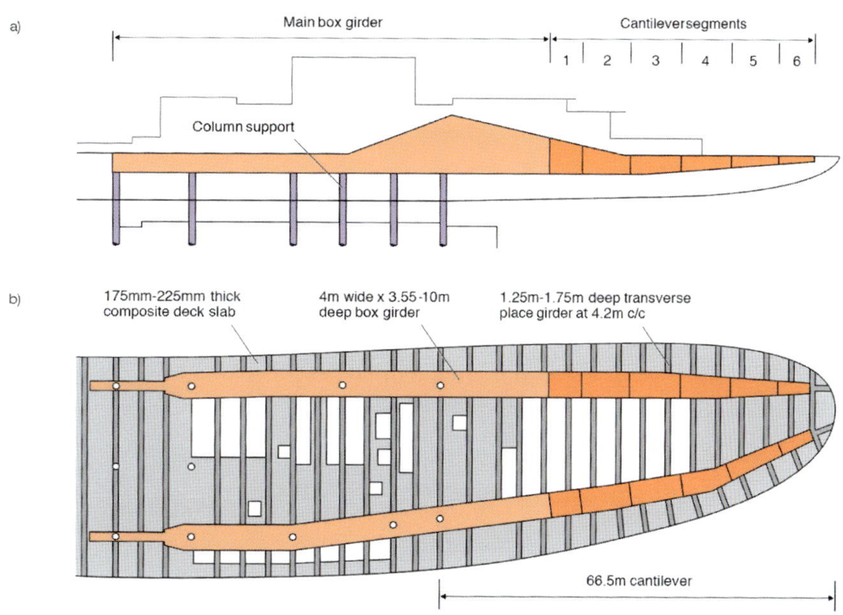

7b6
Cantilever structure
© Nigel Whale/Arup

The hangover

The cantilevering viewing platform is a steel frame split up into six segments with a main box girder backspan. Understanding its behaviour under wind, and the effect of pedestrian-induced vibration, required Arup's building and bridge dynamics experts to work together in an unprecedented way – there are very few structures in the world that have both bridge and building design complexities to this extent. Two stiffened and post-tensioned box girders between 3.55m and 10m deep were designed for the cantilever (Figure 7b6).

Since the cantilever is used by the public, there was concern that people dancing in unison could cause the structure to sway at its natural frequency, causing discomfort to those using the SkyPark. As far-fetched as this may sound,

7b7
Around 120 people at the tip of the cantilever for dynamic testing
© Arup

the phenomenon has occurred on bridges around the world and needed to be designed out. Arup's dynamic, wind and structural engineering specialists worked together to propose that a 5t tuned mass damper should be positioned at the cantilever's tip to control its dynamic response.

The alternative solution to damping was to stiffen the cantilever with additional steelwork, but this would have had the effect of increasing its weight, putting extra strain on the structure or causing more deflection at the tip. The final solution was optimised to include some additional stiffening to complement the function of the tuned mass damper.

On completion of the SkyPark, more than 120 people danced and jumped up and down on the tip of the cantilever, verifying in glorious celebration, the integrity of the final solution (Figure 7b7).

Belly beautiful

To draw attention to the building's showpiece from ground level, Arup façade and structural engineers spent much time on perfecting the soffit cladding. This eye-catching curved underside is made up of separate panels that are arranged to form a geometrical pattern at their joints. Total Design was essential for executing this level of detail, especially where movement had to be accommodated in the bridging sections. The solution was to use springs to fix the gap width between cladding panels, which could adjust for varying degrees of movement. Where the curve at the cantilever tip became more rounded, panels were swapped for a single piece of metal plate, which was profiled to match (Figure 7b4).

Inner sanctuary

Structural, acoustic and façade engineers developed details together to block acoustically weak pathways in the hotel, especially between the SkyPark and guest rooms, many of which sit below the SkyPark's plant rooms. Structural layouts were reconfigured, acoustic buffer zones included or offsets designed in to preserve the sanctuary of these areas and enhance the sensory experience expected by visitors to the luxury hotel.

Raising the roof

Erection of the SkyPark was something Arup considered right from the start of the design process. Design and construction teams, including the structural, civil and bridge engineers, regularly met to discuss lifting procedures to ensure structural members were designed appropriately. These discussions, in the spirit of Total Design, influenced both the SkyPark's and the towers' structure.

Initially, SkyPark sections were to be lifted from the ground using strand jacks on the straight-legged western side of each tower. But this would have required the towers to have significant strengthening to support the overturning forces that would develop. Instead, strand jacks were attached to either side of

7b8

Erection of the box girders (each approximately 700 tonnes) of the SkyPark (below) and cantilever steelwork in situ (left)

Left: © JFE/Yongnam JV Below: © Arup

the top of each tower so that lifting could be more balanced and opposing overturning forces would cancel out. An extension arm was used on the eastern edge to locate the strand jacks beyond the toe of the sloping walls.

A movable lifting gantry (more widely used in bridge construction) was erected on the secondary beams between the main box girders of the cantilever. The gantry was used to lift the six cantilever segments, each weighing approximately 700 tonnes (see Figure 7b8) and was advanced as erection proceeded.

In all, the SkyPark is split into 14 prefabricated steel components, each taking almost a whole day to be carefully lifted into position 200m above the ground. After lifting, they took five days to fix, check and finish.

This ambitious project has been successful due to the exceptionally high levels of collaboration across different disciplines, often across different time zones. Some aspects of the hotel's design were worked on 24 hours a day as Arup engineers in Hong Kong, Singapore, Boston, New York, London, Sydney and Melbourne passed the baton on to the next team at the end of their working day. With engineers moving onto site for key periods or closer to other design offices to create centres of excellence, everything that could physically be done to ensure maximum collaboration and coordination, was done – ensuring that no development of any part of the design would later become an obstacle to achieving optimum results.

A trio of towers

Each 57-storey tower takes the form of a rectangular block with its lower half peeled away to create space for an atrium. These atria continue across the gaps between the towers to create a unified grand atrium at ground level. The "sloping leg" and hence, atrium height and width, is most pronounced for Tower 1, the southernmost tower, where the hotel footprint is widest.

The sloping walls of Towers 1 and 2 imposed significant permanent lateral forces on each structure, exceeding expected temporary wind loads and creating a tendency for the buildings to overturn during construction.

Short-term deformation caused by the self-weight of sloping columns pulling the structure out of line (Figure 7b11) was offset in Arup's structural design by applying a pre-camber to the hotel's walls and floors. This ensured that when the building was complete, these members moved to their correct final positions.

The sloping legs of each tower connect back to the main frame at level 23 via a double-storey-height steel truss (Figure 7b12).

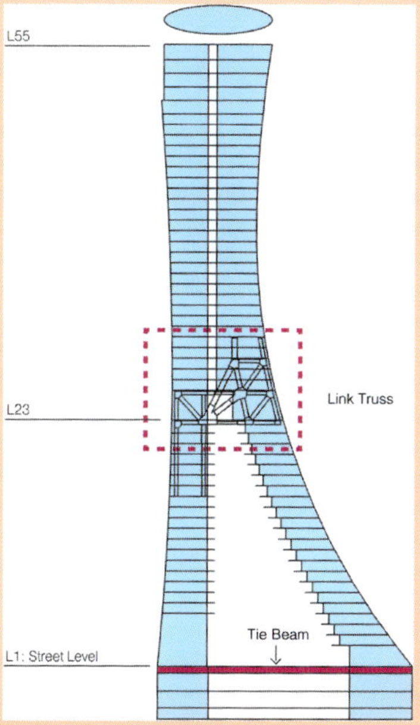

7b9
Section through Tower 1 showing atrium, sloping leg, truss and SkyPark
© Arup

Figure 7b10
The grand atrium
© Paul McMullin

7b11
The tendency for rotation and settlement in Towers 1 and 2
© Nigel Whale/Arup

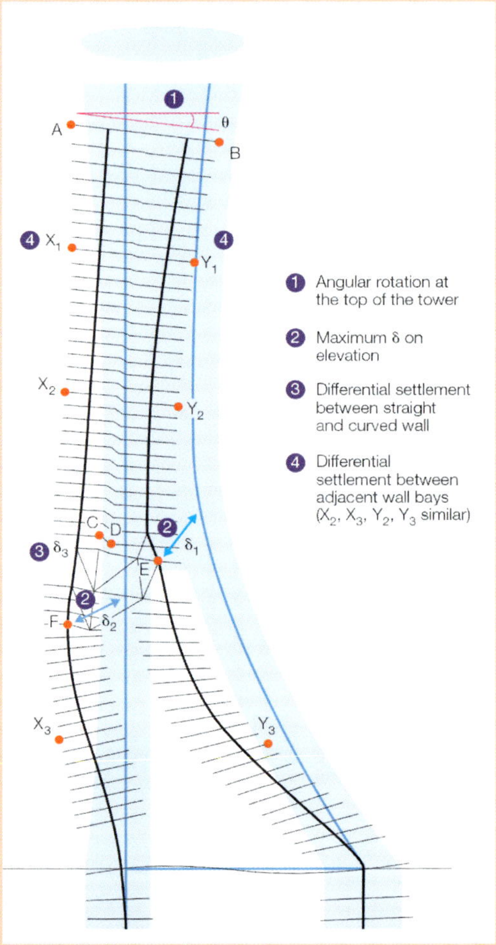

Before the structure had reached level 23, there was concern over deflection in the unstable, part-built structure. To control excessive movement, which could cause a concentration of stresses to develop and the concrete in the walls to crack, post-tensioned tendons were incorporated in key walls, together with a tie beam at level 1 (Figure 7b9).

7b12
The truss at level 23 connects the sloping and straight legs of each tower
© Arup

Largest cofferdams ever used

Underlying Marina Bay Sands' reclaimed land is unstable, soft marine clay. With over 40% of the resort's concrete construction activity taking place 18m to 35m underground and with an average basement depth of 20m, one of the main considerations was to develop a safe and quick method of excavation. Arup, also geotechnical engineer for the project, designed five giant reinforced concrete cofferdams in the ground within which piling and substructure construction could take place, enabling a faster and safer build programme (Figure 7b13). These cofferdams are some of the largest in the world, sinking up to 18m below ground with diameters up to 165m. Their sizes and shapes offered practical advantages to substructure construction since time-consuming temporary propping would not be required.

The five cofferdams plus another diaphragm wall enclosure allowed six separate construction sites to operate simultaneously. Close coordination between geotechnical and structural engineers allowed the walls, in some locations, to also form part of the permanent works.

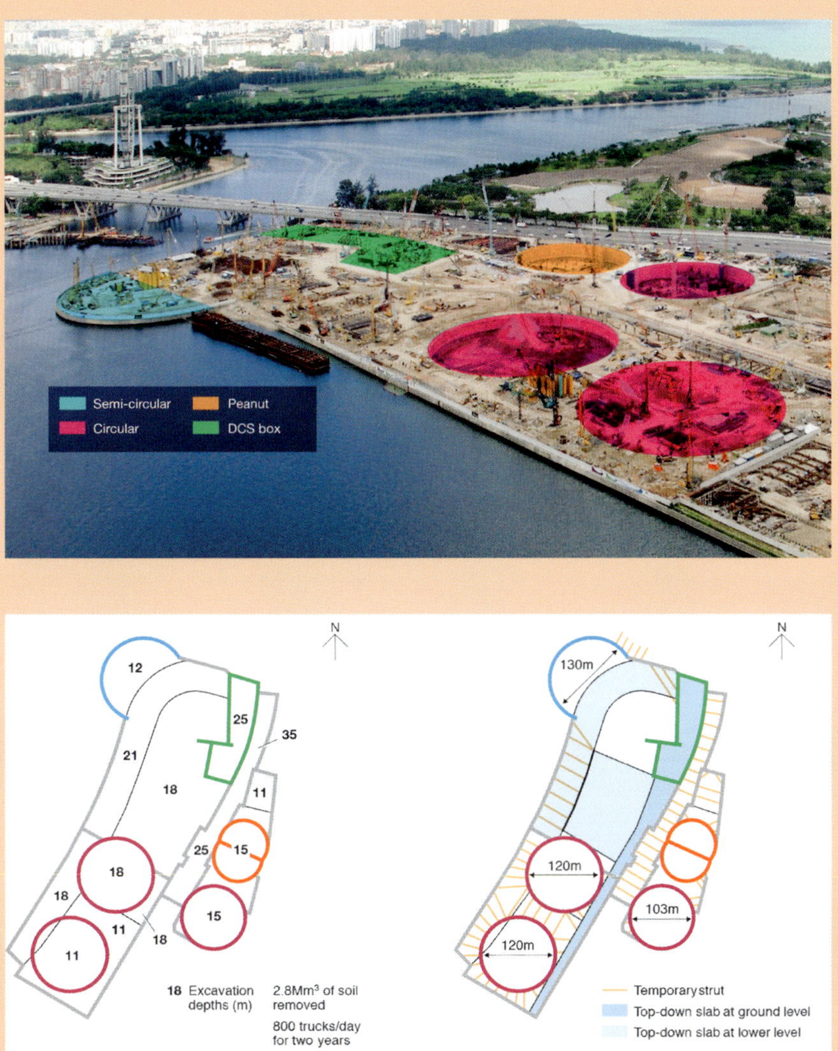

7b13
Aerial and plan views of site showing location of cofferdams and depths of excavation
Top: © Arup;
Bottom: © NigelWhale/Arup

Index